AGRICULTURAL ANIMALS OF THE WORLD

W. Stephen Damron
Oklahoma State University

KENDALL/HUNT PUBLISHING COMPANY
4050 Westmark Drive Dubuque, Iowa 52002

This edition has been printed directly from camera-ready copy.

ISBN 0-7872-4684-0

Printed in the United States of America
10 9 8 7 6 5 4 3 2 1

TABLE OF CONTENTS

Preface

This text is designed to be used with a course designed to interject an international dimension into a curriculum. This may lead the reader to wonder what an international dimension is. An international dimension is nothing more than an element of learning directed at the world or some portion of it other than the nation of the learner, which improves the learner's world view and/or understanding. Providing an international dimension to educational programs in the United States would then include all teaching and research conducted within American universities relative to nations, societies, and cultures outside the United States. This definition includes all teaching and research on topics explicitly treated in a multi-national, comparative, or international manner such as international trade, international relations, international development, and any other comparative studies which involved the inclusion of a foreign element. It would also include study of any collateral area which facilitates the understanding of the topic such as language or cultural studies.

The process of internationalization of the university is the process of including international content, materials, activities and understandings into the teaching, research and public service functions of universities in order to enhance their relevance in an interdependent world (Hathcock, 1994). From the perspective of teaching programs, courses in international trade, economics of agricultural development, crop production in the tropics, and world animal agriculture, are all examples of agriculture courses which qualify as providers of an international dimension. An international dimension can also be provided by including comparative foreign area content in courses not having a specific international focus. This could be in any area including agricultural core courses in any of the disciplines (Kellogg, 1984). Of course, many courses in the arts, humanities and social sciences would fit under either category mentioned above. Obviously, world geography, languages, anthropology, religion, world political systems, and world ecology qualify. From my perspective, any course capable of providing one or more of the major globalizing influencers (cultural awareness; language awareness; geographical awareness; political awareness; climate awareness, and religious awareness) to the students taking them qualifies as having or providing an international dimension.

The purpose of international dimension classes is to offer students a broadened perspective of the world. Specifically, they are intended to help students explore a culture or several cultures which are different from their own. This can be accomplished in several ways but the important link in all of them must be a significant sociological exploration of different societies. International dimension courses offer the opportunity for us to consider our provincialism or narrowness of outlook and expand our horizons. How do we accomplish this in Agricultural Animals of the World? By meeting the following objectives:

1. To examine the roles of animals around the world and how different peoples use them. This includes traditional and contemporary roles. The various relationships of animal, plant and human life are easier to see when we stop considering animals only from our own biased perspectives, and consider broader perspectives and possibilities.

2. To examine the relationships between humans and animals around the world and to demonstrate means for strengthening these relationships. We will be comparing and contrasting the ways of other cultures with our ways as well as offering insight into different cultures, different ways of doing things, and the diverse needs of human societies.

3. To demonstrate that we can better feed people by increasing human, plant and animal relationships. Certain roles of animals as they are used around the world provide incentive for farmers to increase food crop production which will help provide for human food needs. Animal agriculture is actually a key to development of appropriate measures for reduction of erosion, improve soil and water conservation and sustainability of crop production through providing incentives to farmers. We will discover that feeding grains to animals does not cause problems of malnutrition among humans no matter what the current media view is. This is but one of a number of concepts which have become popular in the media but are untrue. Many groups in the U.S. and several European countries are using "simple observations" to arouse emotionalism hoping to influence attitudes against the use of animals. They believe the ways animals are currently used are all unreasonable.

4. For centuries, domesticated animals have provided benefits to the human race in the form of food, clothing, power, soil fertility, fuel and pleasure. In the process they have become integral parts of the economy and/or culture of human societies. Our ultimate objective is to demonstrate that there is every reason to predict that agricultural animals will increase rather than diminish in importance to man.

The numbers, kinds, distribution and utilization of agricultural animals throughout the world and the systems of animal production commonly used are largely determined by these three factors: the physical environment, economic factors and socio-cultural factors. These are the primary influencing factors that we will be considering in this text. The physical environment is primarily beyond man's control, although he can modify it to various degrees. The extent to which the physical environment is modified is directly related to the economic level, ranging from considerable modification in developed societies to none in primitive societies. Economic and socio-cultural factors are interrelated and are the ones largely responsible for the wide variations in animal agriculture in various parts of the world. The animal's ability to adapt and change to these factors will largely determine its success.

Perspective. In modern highly industrialized societies, agricultural animals are a very important part of the economy but their cultural role is largely historical rather than current. Their worth is now almost entirely defined in monetary rather than in aesthetic, social, or cultural terms. At the other extreme, the lower the level of economic development of a society, the greater the role animals play in the culture of that society and the less their worth is apt to be measured in monetary terms. It is very important that the reader understand this difference in perspective. Failure to appreciate the vital importance of this social dimension in the animal agriculture of low subsistence and primitive societies has not only doomed many foreign aid programs to failure but, in some instances, the programs have even been detrimental. It is also the basic reason for most misconceptions about the inefficiencies of animals in the developing countries.

The following set of references were instrumental to me in formulating my ideas about an international dimension course. Some are cited, some not. I reference them as good articles and recommend them for your reading.
3903preface.doc

References

Bonham, G.W. 1989. Education and the World View, in IV. Education and the World View, Council on Learning. Change Magazine Press.

Brandt, K.G. 1987. IX. International agricultural curricular dimensions for the future. in Curricular Innovation for 2005-Planning for the Future of Our Food and Agricultural Sciences. The North Central Curricular Committee Project. P. 69-85. University of Wisconsin-Madison and USDA Higher Education Programs.

Gold, M.A., R.D. Harris, J.R. McKenna and R.L. Young. 1994. Development of graduate and undergraduate programs in agroforestry at Virginia Tech. NACTA Journal 38(2): 21-25.

Hammig, M.D. and C.P. Rosson, III. 1989. Agricultural curriculum: whether an international dimension. NACTA Journal 33(2): 36-39.

Helming Group. 1989. Searching for the ideal agricultural graduate: A survey of the agriculture, agribusiness and food industries. College of Agriculture, University of Nevada, Reno.

Henson, J.B. and Noel, J.C. 1989. Faculty and the Internationalization of the Agricultural Education curriculum for the Year 2005 - Educating for a Global Perspective. The North Central Curricular Committee Project, p.19-25.

Kellogg, E.D. 1984. Providing an international dimension to curricula of agricultural students. NACTA Journal 28 (3): 18-25.

King, D.R. and R.A. Martin. 1994. Infusing a global perspective into the college of agriculture curriculum; topics, activities, and problems. NACTA Journal 38 (2): 39-42.

Kunkel, H.O. 1992. Overview in Board of Agriculture, National Research Council. Agriculture and the Undergraduate. National Academy Press, Washington, D.C.

Martin, R. 1987. Project to Infuse and International Perspective into the Agricultural Education Curriculum. The National Council of Agricultural Education, Alexandria, VA. In: King, D.R. and R.A. Martin. 1994. Infusing a global perspective into the college of agriculture curriculum, topics, activities, and problems. NACTA Journal 38(2): 39-42.

McBreen, Edna L. and Susan G. Schram. 1992. The global context of agriculture in Board of Agriculture, National Research Council. Agriculture and the Undergraduate. National Academy Press, Washington, D.C.

Mason, S.C., K.M. Eskridge, B. Kliewer, G. Bonifas, J. Deprez, C.M. Pallas and M. Meyer. 1994. A survey: student interest and knowledge of international agriculture. NACTA Journal 38 (2): 34-38.

Merritt. R.H. 1984. Challenges for undergraduate education in agricultural sciences. NACTA Journal 28 (3): 9-14.

Rogers, LeRoy. 1994. Impact of long term foreign assignments on professional activities of faculty. NACTA Journal 38 (1): 54-57.

Chapter 1

INTRODUCTION

Food does not come from the grocery store, it comes from the land. Agriculture is what man practices when he tends the land to produce food. Either directly or indirectly, every person on the planet depends on agriculture for its daily food. Because that is true, it is also ultimately true that all of man's other occupations are tied to agriculture. In fact, the entire urban industrial complex of the developed world is sustained only because of food surpluses generated by agriculturists.

The subject of this text is agricultural animals. In it we will be considering how animals provide for man's needs and how they fit into agriculture. This will necessitate a look at the various agricultures and what caused them to develop and be sustained (or fail) in the modern world. We will explore the uses that man has found for his agricultural animals in addition to food and we will individually consider some of the species used by man.

There are approximately 4.1 billion large farm animals distributed throughout the world. (See **Table 1**). That number has been remarkably stable for the last 20 years or more. However, there have been, and continue to be, shifts in the size of individual animal populations. Over two-thirds of the large farm animals are found in developing countries which produce only about one-third each of the meat, milk and wool produced in the world. There are several reasons for this low productivity: unfavorable climatic environments, the prevalence of diseases and parasites, genetically inferior livestock, poor management practices, and sociological factors. To this list must be added another reason that is not generally understood in the developed countries. In most of the developing countries the primary reason for keeping agricultural animals is for something other than food or body coverings. Cultural needs are often the primary reasons for keeping animals. Efforts to improve the agricultural economy and/or level of human nutrition in the developing countries by improving the animals as producers of food or body coverings must also take into consideration the traditional roles of agricultural animals in these societies.

Agricultural animals have always made a major contribution to the welfare of human societies by providing a variety of products and services as shown in **Table 2**. They are a renewable resource, and utilize another renewable resource, plants, to produce these products and services. In return, the manure produced by the animals helps improve soil fertility which benefits the plants.

Animals as Producers of Food

Food is, by far, the most important contribution of agricultural animals, although they rank well behind plants in total quantity of food supplied (**Table 3**). Plants supply over 84% of the total calories consumed in the world, primarily because such a high percentage of the diet in the developing countries is of plant origin. There are several causes of unrest in the poor countries of the world with hunger being one of the most important. As can be seen from the data in **Table 3,** the population of a "poor" (Third World) continent like Africa eats only about 61% of the

Table 1. Livestock Numbers in the World (x1000 head)(1994)

Animal	World Total	Africa	North America	South America	Asia	Europe	Oceania
Cattle	1,288,124	192,180	163,354	279,656	410,118	107,158	34,049
Sheep	1,086,661	208,845	17,529	94,054	340,102	130,692	182,758
Pigs	875,407	21,080	93,056	49,846	482,889	167,983	4,910
Goats	609,488	176,089	14,944	22,819	373,005	14,809	1,007
Buffaloes	148,798	3,200	9	1,436	143,640	133	-
Horses, Asses & Mules	116,882	19,560	20,150	22,049	43,456	5,367	439
Horses	58,158	4,758	12,740	14,493	16,007	4,185	430
Asses	43,772	13,408	3,703	4,105	21,325	906	9
Mules	14,952	1,394	3,707	3,451	6,124	276	-
Camels	18,831	13,815	-	-	4,761	-	-
Total	4,144,194	634,769	309,042	469,860	1,797,971	426,142	223,163
			Poultry Numbers in the World (x 1,000,000)				
Chickens	12,002	976	2,086	1,166	5,553	1,154	85
Turkeys	247	6	100	11	11	85	1
Ducks	681	16	15	13	575	41	1
Total	12,930	998	2,201	1,190	6,139	1,280	87

Source FAO Production Yearbook, 1994
3903chap1.doc.sum96

Table 2. Contributions of Agricultural Animals to Human Societies.

1. Food:

   A. High nutritional value.
   B. Adds flavor and variety to human diets.
   C. Animals convert nonedible materials to food that can be consumed by humans.

2. Body Coverings

   A. Wool
   B. Leather, pelts, hides
   C. Hair, fur

3. Work
   A. Field work and other labor
   B. Transportation

4. Body wastes

   A. Fuel
   B. Fertilizer
   C. Construction material
   D. Feed

5. Miscellaneous

   A. Storage of capital.
   B. Storage of food.
   C. Buffer for fluctuating grain supplies.
   D. Enhance soil fertility.
   E. Income.
   F. Prestige.
   G. Religion and other cultural needs..
   H. Slaughter by-products.
   I. Recreation and sport.
   J. Pest and weed control

Most of the contributions listed in **Table 2** are self explanatory, while others require some explanation. All will be discussed in detail in the appropriate sections in later chapters of this book.

Table 3. Daily Per Capita Intake of Calories and the Percent Supplied by Plants and Animals. (FAO Production Yearbook, 1994)

	Total Calories	Percent Supplied by Plants	Percent Supplied by Animals
World	2718	84.1	15.9
United States	3732	67.1	32.9
Africa	2282	92.6	7.44

daily calories eaten by the "richer" (First World) population of a country like the USA. Since this is an average, it means that there are many people in the Third World Countries who consume far less per day and are severely undernourished, if not on the verge of starvation. Approximately 20% of the population in 90 developing countries are undernourished.

Animals are a more important source of protein than they are of calories, supplying 35% of the protein consumed in the world. Meat, milk and fish are about equal sources of animal protein, supplying, respectively, 35%, 34% and 27% of the world supply of total protein. Eggs are a poor fourth, accounting for only 4.5%.

Animal foods are generally preferred over plant foods by human populations. In fact, one measure of the level of living standards of a country is the proportion of its food supply that consists of animal foods. Not only are animal foods palatable and delicious, they are also the most nutritionally complete foods known. They are an important source of vitamins and minerals, and the protein in animal foods are more likely than are plant proteins to include the essential amino acids in the correct proportions.

There are many who feel that because the world population is growing at a faster rate than is the food supply, we are becoming less and less able to afford animal foods because feeding plant products to animals is an inefficient use of potential human food. It is true that it is more efficient for humans to eat plant products directly rather than to allow animals to convert them to human food. At best, animals only produce one pound or less of human food for each three pounds of plants eaten. However, this inefficiency only applies to those plants and plant products that the human can utilize. The fact is that over two-thirds of the feed fed to animals consists of substances that are either undesirable or completely unsuited for human food. Thus, by their ability to convert inedible plant materials to human food, animals not only do not compete with the human for food, rather they aid greatly in improving both the quantity and the quality of the diets of human societies.

Table 4 presents some statistics that are ignored by those who would suggest that we can no longer afford the luxury of animal foods. Only about 35% of the land area of the world is classified as agricultural. Thus, roughly two-thirds of the land area of the world is not suited for any sort of agricultural use because it is covered by cities, mountains, deserts, swamps, snow, etc. Of the 35 % that can be devoted to agriculture, less than one-third (or about 10% of the total land area) can be cultivated and produce plant products that the human can digest. The remaining two-thirds of the world's agricultural land is covered by grass, shrubs or other plants that only ruminant animals can digest. Thus, the inefficiency of animals is not a major concern since they represent the only way these plants can be converted to human food. As the human population of the world increases, it is likely that we will be forced to depend more and more on animals, especially ruminants, to meet the increased demands for food.

As one studies **Table 4**, there are several things that stand out. The developed countries of the world, that is the richer countries, are not that way because most of their land is agricultural land. The developing (poorer) countries have about the same percentage of their total land area that is classified as agricultural. It is true, however, that the two most highly developed

Table 4. Characteristics of Agricultural Land in Various Geographical Regions.

Geographical Region	Total Land Area	% of Total Land that is Agricultural	% of Agricultural Land that is Cultivated Land	% of Agricultural Land that is Permanent Pastures
	(1000 sq. mi.)	(%)	(%)	(%)
World	50,495	35	31	67
Developed Countries	21,176	36	33	66
Developing Countries	29,319	34	29	69
Africa	8,994	37	19	79
Asia	10,334	38	45	53
Europe	1,826	49	55	38
Oceania	3,254	61	9	91
N. America	7,084	27	46	53
S. America	6,771	31	15	81
USSR	8,599	27	38	62
USA	3,524	47	43	56

Source: FAO Production Yearbook.

continents (Europe and N. America) have more than the average amount of cultivated land. Oceania (primarily consisting of Australia and New Zealand) ranks first in total agricultural land (61%) but a poor last in amount of cultivated land (9%).

The United States, and other developed countries, feed a considerable amount of grain to animals, even to ruminants that have the ability to subsist entirely on roughage materials that the human cannot digest. There is only one reason why this is done and that is because it is currently profitable to do so. In this way ruminants, and other agricultural animals, serve an important function in serving as a buffer for fluctuating grain supplies. Feeding surplus grain keeps the price farmers receive for their grain at a higher price than it would be if the grain was not fed. However, if a shortage of grain, whatever the reason, causes the price of grain to rise sufficiently to make it unprofitable to feed to ruminants the practice would cease. However, should this happen, ruminant animals would still be needed for their primary function of converting inedible plant materials to edible human food.

Monogastric animals like swine and poultry can be competitors with the human for food if they are produced by the intensive confinement systems widely practiced in the developed countries. In fact, the highest proportion of feed grains and other concentrates, such as oilseed meals, fed to livestock in the United States are fed to swine and poultry. As is the case for ruminants, this is done because it is profitable to do so. Likewise, the feeding of grain will be stopped if the demand for grain for human consumption forces the price too high. In this case, unlike the situation with ruminants, the numbers of swine and poultry will probably be reduced somewhat in the developed countries. However, they will not disappear as an agricultural animal because, as will be discussed in a later section, they have a unique ability to convert human garbage to human food. This is one of the most important functions of swine in agriculture today.

CONCLUSION

Agricultural animals serve a wide variety of needs for man. Animals help us use our resources to feed our population. While agricultural animals have come under attack in recent years by those who feel they are a luxury, agricultural animals are not destined to disappear, or even be reduced in numbers and importance. What is most likely to happen is they will play an even more important role than they have in the past in helping alleviate human hunger. For this reason, it is important that various people learn something about the factors that determine the kinds of agricultural animals found throughout the world.

References

CAST. 1980. Food from Animals: Quantity, Quality and Safety. Report No. 82, Council for Agricultural Science and Technology, 250 Memorial Union, Ames, IA 50010.

Jordan, T.G., M. Domesh and L. Rowntree. 1994. The Human Mosaic. 6th ed. Harper Collins College Publishers

FAO. 1994. FAO Production Yearbook, FAO, Rome.

McDowell, R.E. 1991. A Partnership for Humans and Animals. Kinnic Publishers. Kinnickinnic Agri-Sultants, Inc., Raleigh, N.C.

McDowell, R.E. 1984. The need to know about animals. In: World Food Issues, M. Droskoff, ed. Center for Analyses of World Food Issues, Cornell University, Ithaca, New York.

Rockefeller Foundation. 1976. The Role of Animals in the World Food Situation. The Rockefeller Foundation, 1133 Avenue of the Americas, New York, NY 10036.

Taylor, R.E. 1995. Scientific Farm Animal Production. Prentice Hall, Englewood Cliffs, N.J. 07632

Study Questions

NOTE: In answering the study questions in this and all later chapters, where fractions are called for or needed in the answer, use the closest fraction. For example if the exact figure is 31.1% the closest fraction would be 1/3. Round percentage figures to the nearest whole percentage.

1. Explain what is meant by the statement that animals serve as "a buffer for fluctuating grain supplies."

2. Plants supply approximately what percent of the total calories consumed in the world?

3. Animals supply what fraction of the total protein consumed in the world? Which 3 animal food products supply most of the animal protein?

4. Approximately what fraction of the total land mass of the world is classified as agricultural land? What fraction of the agricultural land of the world is not under cultivation and is primarily devoted to grasslands?

5. Give two examples of how animals convert inedible materials to human food.

Chapter 2

Adaptation

Introduction

When one considers the many factors that are concerned in shaping the kinds of animals, both wild and domesticated, found throughout the world, there is none more important than the ability of the animal to adapt to the specific stresses of the environment in which it must live. The ability to adapt to a symbiotic relationship with humans and the stresses that association imposes is a key element of domestication. In this chapter we will take a look at what adaptation is, and look at the stresses to which an animal must adapt. In following chapters we will discuss those stresses in greater detail.

Adaptation is the sum of the adjustments occurring in an organism that promotes its welfare and favors its survival in a specific environment. It should be easy to appreciate that these adjustments must be to the unfavorable features, or stresses, of the environment.

The **environment** is all of the combinations of conditions under which an organism must live. There are many of these but they can be classified into two groups:

(1) The natural, or physical environment.
(2) The artificial environment imposed by man.

The stresses of the physical environment may be classified into five categories: (1) climatic stresses; (2) nutritional stresses; (3) internal stresses; (4) geographical stresses; and (5) social stresses. Each affects the distribution and utilization of the agricultural animals of the world. A brief discussion of each follows.

Climatic Stresses: Climate is the longtime change in certain meteorological factors, while weather is the immediate condition of these factors in a given area. The most important of these meteorological factors from the standpoint of animal comfort and performance are ambient temperature, rainfall, solar radiation, wind and relative humidity. Each has both a direct and an indirect effect on animals, with the indirect effect primarily through their effect on plants that the animal must use for food. Generally speaking, the climate and weather of an area are the same. However, they may differ temporarily, such as when an area with a rainy climate suffers a drought.

Nutritional Stresses: These are the stresses related to the quantity and quality of available feedstuffs. In most cases they are determined primarily by the climate of an area, with temperature and rainfall being the most important. However, they are also related to soil type and fertility, topography and geographical location.

Internal Stresses: These are the stresses that affect an individual animal by gaining entrance into the body. The two most common ones are pathogenic organisms that are the cause of disease, and toxic substances and materials that may be found in a given area.

Geographical Stresses: These are the non-climatic stresses associated with a particular geographical location. The most important of these are the stresses of high altitudes or the stresses associated with the type of terrain. Examples of the latter include: constant snow or ice cover; soft, shifting terrain such as sand; soft, wet terrain such as swampy areas; steep areas; rocky areas, etc.

Social Stresses: These are stresses associated with the interactions of the animal with other animals, including humans. It should come as no surprise that the most important stresses in this group affecting domesticated animals are those associated with the human. However, man also has an increasingly important effect on the wild species. Interactions with other animals includes those of the same species, in which competition for the available space and food may be a major problem, and with animals of other species. An important example of the latter is the predator-prey relationship.

Artificial environments are created by man to make the animals best serve him. One part of the artificial environment is the physical environment as modified by man. For example, he can irrigate very dry areas or provide shelter for his livestock in very cold areas. He can carry out disease prevention measures to reduce the detrimental effects of a disease environment. He can also impose management practices that increase the detrimental effects of the physical environment, such as confining animals and not permitting them to seek shelter from unfavorable conditions of the physical environment, or increase their exposure to diseases and parasites by overcrowding.

The specific nature of the artificial environment imposed by man is determined by the nature of the "service" demanded from the animals. In developed countries, such as the United States and Europe, this is usually the maximum production of some product useful to man, with the most common being food or some type of body covering. In developing countries, the nature of the service is usually quite different such as work, transportation or religion with food sometimes being much less important. Generally speaking, changes of this sort to benefit man are not beneficial to the animals, and usually reduces their ability to survive in the wild. (We'll discuss this in a later section.)

Because the major emphasis differs between developed countries, and developing, it should not be surprising that animals well adapted to the artificial environment of the developed countries would not be equally well adapted to the artificial environment imposed by the cultural and economic factors of the poorer nations of the world.

How do animals change in response to the environment? The adjustments (changes) that occur in individual animals to make them adapted to specific environments are of three types: (1) **morphological or anatomical changes**, that is, changes in form and structure. These changes are external in nature and easily seen. We will consider a number of these including, for the sake of example, changes in quantity, quality and nature of body coverings and body appendages, difference in color, presence and absence of body parts such as horns, humps, etc.; (2) **physiological changes**, which are primarily changes in the biochemistry of the body. These are internal changes and not often readily apparent. These are more profound changes and are more difficult for the animal to acquire; and (3) **behavioral changes**, or "learned" responses. These involve a number of things, with probably the most common being those associated with self-protection and seeking food and water.

Acclimatization may be defined as the adjustments that result in increased tolerance to climatic stresses. Thus, acclimatization is adaptation to climatic stresses, but it differs from adaptation in that adaptation represents adjustments to all stresses. Frequently the word acclimatization is used when the correct term is adaptation. For example, the statement is often made that it is important that animals acclimate to an area. This is correct if the only stresses present are climatic. However, if other stresses are present, to be correct the statement should use the word "adapt" rather than "acclimate."

CONCLUSION

Agricultural animals must adapt to a wide variety of stresses if they are to be of use to man. In this chapter we defined adaptation and briefly described the environmental stresses to which an animal must adapt. In the next 2 chapters we will consider in greater detail the profound influence that some of these stresses have. Specifically we will discuss climate and social stresses. The other stresses are important but more so to animal caretakers and are less essential to the broader understanding of the concept of adaptation. Understanding them also requires a good deal of understanding of the physiology of the body. For these reasons we will not discuss them in great detail.
3903chap2A.doc

Study Questions

1. What is adaptation?
2. What is the difference between the natural and the artificial environment?
3. How does climate differ from weather?
4. Explain why adapted animals are acclimated but acclimated animals are not necessarily adapted.
5. List and briefly describe the 5 stresses of the physical environment under which organisms must live.

Chapter 3

The Adaptation of Animals to the Economic and Cultural Environment of Human Societies

Introduction

Chapter 2 pointed out that agricultural animals must have the ability to adapt to the artificial environment imposed by man if they are to be useful to him. One part of this artificial environment is that associated with the economic level and the culture of human societies. The combined social and cultural influences man has imposed upon himself have a tremendous influence on what people of various societies eat and thus they affect animal agriculture and animal use throughout the world. This allows religion, along with recreation and social custom to be very important in establishing an animal's value. A value established this way is often greater than the animal's food or work value.

Many social and cultural influences are region and/or people specific, though certainly not all. They are also dynamic and evolving and influenced by the economic situation. An example is the animal rightist movement. This is an issue only in wealthy Western societies not in countries less fortunate. It is easier for people who aren't hungry to be concerned over such matters. However, regardless of the issue or the motivation, social and cultural structure have tremendous influence on what specific peoples eat. Eating is seldom done simply to satisfy hunger or fulfill nutritional needs, even though nourishment is one of man's basic needs.

Cultural and economic levels are both considered in this chapter because it is often difficult to clearly separate the two. It is also difficult to clearly define their separate effects on the number and kinds of agricultural animals that are found in a given area of the world and how they are produced and utilized by the human inhabitants of this area.

Religion

Religion is an important part of the culture of human societies. **Figure 1** presents the worldwide distribution of the major religions of the world. The approximate numbers of adherents of each religion are given in **Table 1.** It is not our intent to discuss in detail all the social and cultural issues surrounding food. However, religion is one part of culture worth discussing in more detail because of the magnitude of its influence on animal agriculture. Even then, we will not discuss all the religious laws which affect animal uses, even though many are fascinating. For example, followers of Jainism frequently wear masks to prevent the inhaling and killing of an insect. Animal slaughter and meat consumption is totally unacceptable. Buddhists, Seven-Day Adventists and Mormons all have restrictive dietary laws regarding meat and other animal products. However, we will not include any extensive information on these for a simple and pragmatic reason: they do not dominate any major country to the degree that they have influenced animal distribution on a major geographic scale. For the most part, we will restrict our discussion to major religions that influence animal agriculture. We will let the following

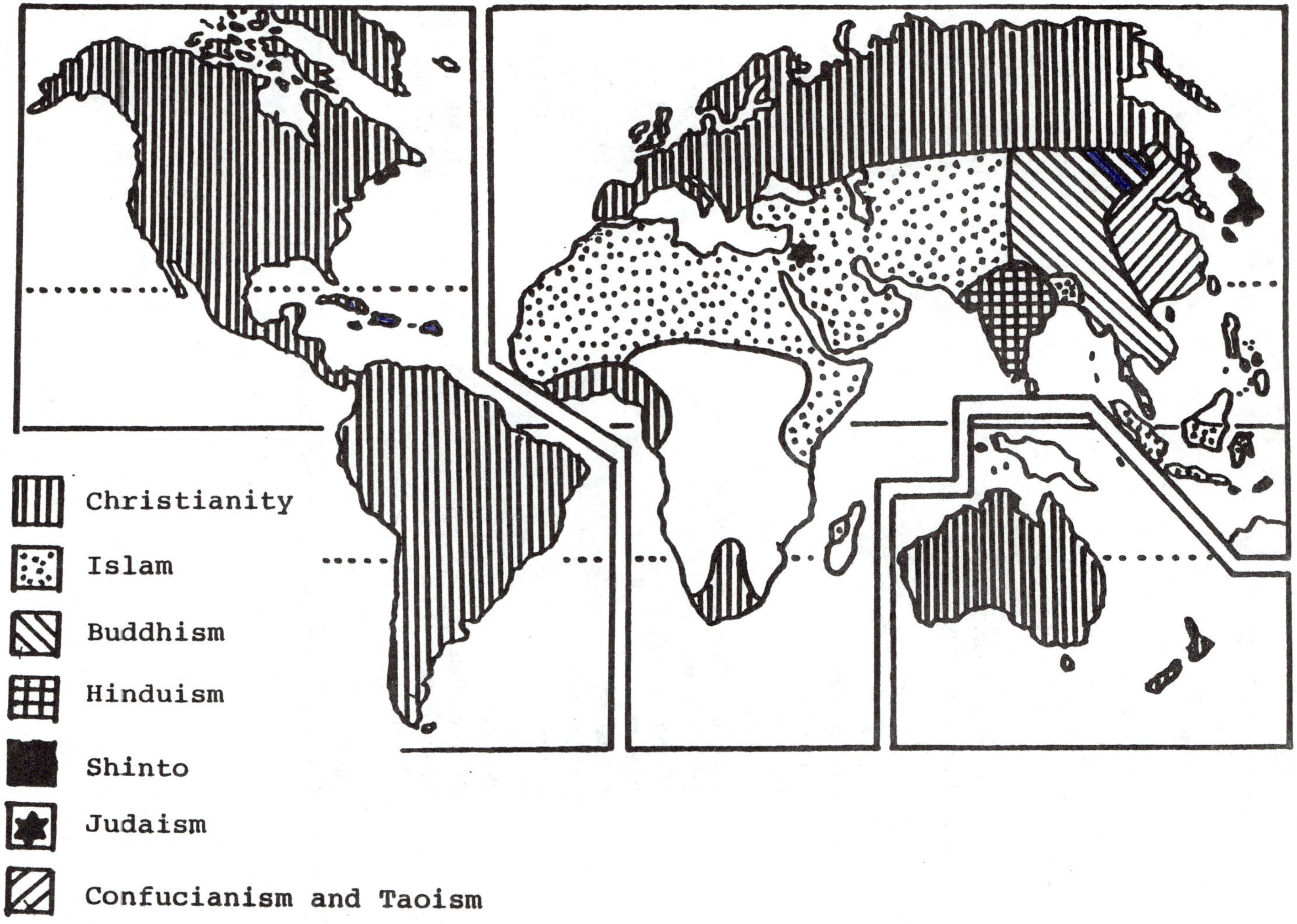

Figure 1. Worldwide Distribution of the Major Religions of the World

Table 1
MAJOR RELIGIONS OF PEOPLE POPULATING THE WORLD

RELIGION	FOLLOWERS
Christianity	
Roman Catholicism	1,042,501,000
Protestantism	382,374,000
Orthodox Christianity	173,560,000
Anglicanism	75,847,000
Other Christianity	195,470,000
Total Christianity	1,869,751,000
Islam	
Sunnite	841,928,760
Shi'ite	162,299,520
Other Islam	10,143,720
Total Islam	1,014,372,000
Hinduism	751,360,000
Buddhism	334,002,000
Traditional Chinese folk Religion	140,956,000
Tribal religions	99,736,000
Sikhism	19,853,000
Judaism	18,153,000
Shamanism	10,854,000
Confucianism	6,230,000
Bahaism	5,742,000
Jainism	3,927,000
Shintoism	3,336,000
Other religions	142,948,000
Non-religious	912,874,000
Atheism	241,852,000
Total	5,575,954,000

Source: Compton's Interactive Encyclopedia. 1995. NewMedia, Inc.

example of Amish agriculture represent and illustrate for the rest just how profoundly religion can affect man's agriculture.

Amish agriculture. An excellent example of how religion can radically affect the agriculture of a people is found with the Amish farmers in the United States. The following excerpt from Jordan, et al, 1994, p.123 is quite instructive on this point.

> "Perhaps no more striking example of the interplay of culture and agriculture can be found than among the Amish farmers of Pennsylvania and the Midwest. Their religious beliefs instruct them not to depend upon nonbelievers for help, leading most to shun modern machines and energy sources. As a result, the Amish rely much less upon inanimate power and agricultural machines than does the average American farmer. A recent study by Warren Johnson and others comparing Amish and non-Amish farmers in three states–Pennsylvania, Illinois, and Wisconsin—not only revealed these fundamental differences in farming practices but suggested that, in some regions at least, the Amish surpassed other agriculturists in the amount of food calories produced per unit of energy expended in the production process. Because of their skills in energy conservation, the Amish might eventually be able to instruct a profligate nation on how to live on less electricity, oil, and gas. In the meantime, the Amish will continue to farm differently because of their religion, classically illustrating the workings of cultural integration."

Three of the great religions of the world have a direct and marked effect on the numbers and utilization of livestock. Islam, whose followers are called Muslims, forbids all contact with swine. As a result, in Muslim areas (primarily North Africa and the Near East) hogs are completely absent. Judaism (Jewish) also considers pork unclean and forbids consumption of pork. Likewise, there are very few hogs in Israel. The Hindu religion is primarily found in India, and is the state religion. About 90% of the population are Hindus. The cow is considered sacred and can neither be slaughtered nor sold for slaughter. As a result there is a vast accumulation of cattle in India which directly contributes only milk to the nutritional needs of the country. This is unfortunate since India has a vast population (919 million in 1994 and increasing at the rate of 17 million per year) over half of which are very poor and undernourished. However, milk is a major food contribution and they make other indirect contributions to the human population as well. Let's explore these contributions further.

The Sacred Cow of India. Although their direct contribution to the food supply of India is hampered because of religious beliefs, the vast numbers of cows do make a major contribution to the country in addition to being a part of their religion. In the eyes of the Indian farmer, the cows require little in the way of care and feeding. They can graze the common lands at no cost, even though the vast numbers create a severe overgrazing problem. In addition, they are needed to consume the vast quantities of straw that are available, since this is about the only means available to utilize the straw. Even though the soil could benefit greatly from added organic matter, the animal-drawn plows they have are not capable of turning straw under and even if they could the multiple cropping that is practiced gives little time for the straw to breakdown. They make a major indirect contribution in the form of bull calves to be used for work oxen. Each farm, regardless of its size, requires at least two work oxen. There are two reasons for this: (1) the animals are underfed so they cannot work hard all day but must be relieved for part of the day; (2) There must be an animal available for use exactly when needed. When the monsoons begin the fields must be worked. For most of Asia there is one draft animal per hectare of cultivated land.

An additional very valuable product contributed by the cows is manure. Unfortunately, as badly as it is needed to enhance soil fertility, much of it is dried, perhaps mixed with more straw, and then used for fuel for fires. There is no good estimate available as to how much is so used, but Harris (1990, p.162) estimates that in India, more than 350 million tons of manure is used annually for home fuel needs. This is done because the people have no real alternative. In the populated areas, the trees were cut long ago and no other fuels are available. India is woefully lacking in natural reserves of fossil fuels. Therefore, cows' manure is dried for fuel for fires. It is considered an excellent source of cooking fire because it burns with a clean, slow, long-lasting flame that doesn't burn the food. The housewife can thus cook as she works and doesn't have to stir so frequently. She can even help in the fields. An equivalent amount of the manure is returned to the land as fertilizer to aid in the production of food from plants.

There is also another side to this issue of meat eating in India. In times of extreme hardship, Hindus have been known to slaughter and eat cattle rather than starve. There are also numerous reports from various venues that indicate that some upper caste Hindus have developed a taste for and do consume at least some beef. In addition, the lower caste "untouchables" are permitted to dispose of dead animals. It would appear that a major method of disposal is to eat them. These same people use the leather in a huge leathercraft industry. Object lesson: **things are almost never as they seem.**

Food and Religion, a general discussion. Many modern day taboos regarding food and the way it is served were brought from ancient to modern times by religion and the rituals and customs associated with religion. In some religions (cultures), this may simply be blessing food to give it special powers. In others, religious law dictates selection and judgment of the quality of food before the gods will approve. In the past, these religious dictates often had a practical basis. It seems much more than coincidence that religious food laws frequently reflect a common sense approach to the prevention of food contamination and food-borne disease. For instance, Kosher food inspection and certification (Jewish) was with us long before modern food inspection. Many other examples could be cited where religious leaders have affected the relationship of man to animals. Sometimes these have had profound long-term influence. Let's use China and India as examples.

High per-capita beef consumption has always been associated with low-density populations who have access to grasslands. These conditions fit the people of ancient northern India, and the ancestral Chinese of the Yellow River basin. In ancient times, both raised cattle and ate beef. The populations grew in both areas, and farming intensified. Cattle became too expensive in both places to be eaten because they were needed as draft animals to produce crops. In China, pork gradually became the preferred meat. This was probably because pigs could be raised more efficiently on household wastes and agricultural by-products than cattle. Cattle were used for draft, if at all. Gradually their numbers diminished. Religion didn't get involved in the process too much. In India, a different course was followed. The fierce summer heat, the monsoons, and periodic droughts were not good conditions for hogs and they didn't flourish. So, as population density increased India turned to milk as the major source of animal protein. It soon became a religious principle of Hinduism and subsequently Buddhism (which sprang from Hinduism) to protect the supply of draft animals by prohibiting the slaughter of cattle. The Hindus carried this to the greatest extreme and a civilization arose in which beef-eating is simply not practiced by the faithful (supposedly).

Examples can be found where a mistake or two seem to have been made (from a practical perspective that is) in the name of religion. While there is logic in Hinduism's protection of the milk and work animals with a taboo on beef slaughter, most Hindus are also forbidden from eating fish. Even though India is surrounded by fish-containing oceans, protein and general food deficiencies are a way of life in India because the population will not consume fish. Seven-Day Adventists also have a finless fish taboo and a taboo against pork. After their missionaries converted the Pitcairn islanders, the self-sufficient economy of the island disintegrated. Prior to their conversion they had depended on pork and finless fish in their diet. (Jordan, et al, 1994, p.238-239).

Values. Ours vs. theirs. (Object lesson) One's own religious beliefs make it difficult to fully understand and appreciate the beliefs of others, particularly if they differ radically. The important thing to remember is that there is no chance that major segments of any religion will change their religious laws. Thus, the situations that exist are the situations that are likely to continue. Whether you or I think it is right is beside the point. (One of the great lessons of life is to learn to graciously accept that you've just been told your opinion is irrelevant.)

Levels of Economic Development

Two terms, D*eveloped countries* and D*eveloping countries* have already been used and will be used repeatedly in this book. They refer to the level of economic development of a country and, in this book, the classification of countries into one or the other will be the same as that used by FAO (Food and Agricultural Organization of the United Nations with headquarters in Rome). Their classification has historically been presented in detail in the preface of each edition of the FAO Production Yearbook, and is summarized in **Table 2.**

Although not used by FAO, three other classifications in common usage which may be encountered are *First World, Second World* and *Third World.* First World refers only to the developed countries with a market economy while Third World are the developing countries with a market economy. Second World Countries are all countries with a centrally planned economy, that is the communist countries, regardless of whether they are classified as developed or developing. This term has been fading from use since the fall of the Berlin Wall. Another designation often encountered is to refer to the developing countries as the *LDC's (Lesser Developed Countries).* There seems to be some confusion as to how to use this term. It seems

Table 2
Developed Countries and Developing Countries:

DEVELOPED COUNTRIES

North America: Canada and United States
Europe
Oceania: Australia and New Zealand
Asia: Israel and Japan
Africa: South Africa
Russian Federation

DEVELOPING COUNTRIES

Africa - All countries except South Africa
Asia except for Israel and Japan
Latin America - All countries of Central America and South America
North America - Bermuda, Greenland, St. Pieree, Miquelong and Mexico
Oceania - All islands except Australia and New Zealand

that it is often used for South American countries and Mexico. They are clearly more developed than most of Africa and could use an additional designation.

As the name implies, the developed countries have reached a higher level of economic development than have the developing countries. The same is true of the countries with a market economy as compared to those with a centrally planned economy. There is also considerable variation in the level of economic development of different nations within each category, particularly among the developing nations. Third World Countries range from those that are poor but which enjoy a fairly high standard of living to the extremely poor countries where per capita income is less than $100 per year.

There are several important reasons for a course concerned with international agriculture. One of the most important of these reasons is to make students aware of the vast differences that exist between the haves (developed countries) and the have nots (developing countries) of the world and the vital necessity to do everything possible to narrow the gap in economic well being between them. One could list many reasons why this should be done, with humanitarian concern heading the list. People fortunate enough to live in a developed country are in a definite minority by a factor of 3 to 1. Poor people are often desperate people with little to lose and everything to gain by rebellion.

Levels of Agricultural Development

In general terms there are three levels of agricultural development: (1)developed; (2) subsistence; and (3) primitive. The livestock systems in a developed agriculture may be further subdivided into: (1) intensive, usually small farm units; or (2) extensive, usually range units with the type and breed(s) of livestock that are most popular determined by whether the range area is humid or dry.

A **developed agriculture** is usually associated with the following characteristics:

- Very few (usually less than ten percent) of the total population on the farm actively engaged in farming.

- A highly specialized agriculture with each unit producing only one or two products.

- A highly mechanized agriculture with little, or no, animal or hand labor.

- A high per capita income.

- A high literacy rate for the total population.

Almost all developed countries also have a developed agriculture. The exceptions are some of the nations in Eastern Europe, and the Russian Federation. Several of these latter two groups of countries are marginal between having a developed and a subsistence agriculture. There are no countries classified by FAO as being a developing country that have a developed agriculture.

Subsistence agriculture is usually associated with:

- Approximately one-half of the total population engaged in farming.

- Each farm family produces roughly what it consumes with only a little surplus for sale or barter. This surplus is barely enough to supply the needs of the portion of the population that is not on the farm producing their own food. Thus, the country has little, or no, agricultural products for export.

- Little mechanization and much hand and animal labor required.

- A relatively low per capita income.

- A relatively low literacy rate.

This is the level of agricultural development found in most of the developing countries outside of Africa.

Primitive agriculture is characterized by:

- Almost the entire population involved in producing their own food since no one produces a surplus that they could purchase.

- Generally a scarcity of food and a low nutritional level in the country.

- No mechanization and very little animal power utilized in farming. Almost all of the labor is hand labor which means that only a small acreage can be farmed by one individual.

- Extremely low per capita income.

- Very few literate individuals in the country.

This is the type of agriculture found in the most backwards of the developing countries. Most of the primitive agriculture is found in Africa.

Table 3 presents a comparison of the percent of the population in various levels of economic development that are engaged in agriculture. Generally speaking, the developed countries have a developed agriculture and the developing countries have either subsistence or primitive.

There are several things that stand out in **Table 3** relative to the percent of the population engaged in agriculture. Farmers are in a definite minority in the United States and other

Table 3

Percent of Economically Active Population Engaged in Agriculture.

Country/Continent	% of Population Engaged in Agriculture
All Developed Countries[1]	8.7
First World[1,2]	5.7
United States	2.0
Canada	2.7
Denmark	3.9
France	4.3
Australia	4.4
Greece	22.0
Albania	45.7
South Africa	12.4
All Developing Countries[1]	60.2
Africa	60.7
Rwanda	90.6
Cuba	17.6
Argentina	9.4
Brazil	21.9
Asia	57.3
Bangladesh	65.8
Bhutan	90.0
China	64.4
World	43.4

Source: 1994 FAO Production Yearbook

[1]Estimated from historical data in FAO Production Yearbooks. In 1993, FAO stopped publishing these statistics in their annual yearbook.

[2]Selected countries used for comparison.

developed countries. Although most developed countries have less than 15 percent of their population engaged in agriculture, there are several that have considerably more (FAO sets 30 percent as the upper limit for classification as a developed country). Included in the group that have more than 15 percent of their population engaged in agriculture are a number of the countries of Eastern and Southeastern Europe that have some subsistence agriculture practiced in part of their country. Because of the political turmoil in that part of the world current statistics are difficult to find and/or trust. Earlier it was stated that in primitive agriculture almost all of the population is engaged in farming. **Table 3** does not appear to completely support this statement. The overall average for Africa, which has most of the primitive agriculture in 68.8 percent, which is hardly "almost all of the population." However, there are many African countries that have 80 to 90 percent of their population engaged in agriculture. Since it is hardly likely that there is any country in which the agriculture is entirely primitive, the 80 - 90 percent suggests that in the areas where primitive agriculture is practiced the entire population is involved.

Literacy refers only to the ability to read and write and indicates nothing about native intelligence or the ability to learn. However, it is a definite handicap to a nation to have a high level of illiteracy in its population. It is one of the major obstacles to efforts to introduce new technology in an effort to improve level of agricultural development. If people cannot read, the only way they can learn of new technology is by being told. An Extension Specialist can write agricultural bulletins that can reach hundreds of thousands of people to tell them of improved agricultural methods. However, if it requires personal contact and verbal explanation, that same Extension worker can reach only a few hundred people per month. Most developing countries realize that the first step to improving their level of agricultural and economic development is to improve the educational level of their population and this is the main reason Developed-Country Universities have so many students from developing countries.

Economic Institutions and Agricultural Development

In countries with a developed agriculture a number of essential economic institutions are highly developed and functional. In contrast, these institutions are relatively poorly developed, if present at all, in countries with subsistence agriculture and are almost totally absent in countries with a high incidence of primitive agriculture. These economic institutions are (1) Financial Institutions; (2) Marketing Agencies; (3) Industrial Institutions; and (4) Governmental Agencies. The following is a very brief discussion of their importance.

Financial Institutions: Included in this group are all of the various types of institutions that are concerned with handling money. They make the following essential contributions: (1) They are a source of loans to finance agricultural enterprises. (2) They provide a place for the investment of the profits from agriculture that are not reinvested in agriculture. (3) They serve as a safe repository of money and provide a means for the safe and orderly transfer of money in financial transactions. The importance of these functions to the economic development of a country's agriculture should be obvious. One of the undesirable side effects of the lack of banks and other financial institutions in the developing countries is the use of livestock, particularly cattle for "storage of capital." This will be discussed in a later chapter concerned with the utilization of agricultural animals.

Marketing Agencies: These are essential to provide a means for those engaged in agriculture to sell their products at a maximum price and at a time they want to sell them. Included are such agencies as local auction sales barns, or large terminal markets. In most of the poorer countries, farmers do not have access to such competitive markets and are at the mercy of local traders who offer very low prices. Marketing agencies also provide the means by which those engaged in agriculture can purchase the goods they need or think they need.

Industrial Institutions: It is impossible for a nation to reach a high level of agricultural development without also having a high level of industrial development. There are three important reasons why this is true: (1) Industrial institutions, such as factories, provide non-farm employment for the 70-98 percent of the working population that are not engaged in agriculture; (2) The wages paid these workers provides a cash market for agricultural products; and (3) These industrial workers produce the consumer goods needed in agriculture, either as essentials or as luxuries. The availability of these consumer goods is also a stimulus to high agricultural productivity to make more money in order to purchase them.

Governmental Agencies: This includes the agencies responsible for transportation (road building and maintenance, etc.), education, protection (both police for internal and an army for external protection), equitable land policies and sound money. Unfortunately, too often the people who have the power to make changes in government to improve these services are unwilling to do so. They profit from the fact that the masses are illiterate or lack good markets. They may be large land holders so would hardly favor land redistribution. In most of the developing countries the task is formidable and looks almost hopeless of solution in the near future. It will require a generation growing up with greatly improved educational opportunities to take over and push for the needed changes.

Conclusion

Man and his domestic animals and improved varieties of plants are influenced by a wide variety of factors. Among those things are cultural issues and the economic development of the people. In this chapter we briefly examined religion as an important issue and we looked at the levels of economic development of various peoples and how these have influenced the different levels of agricultural development. We also discussed the world's economic institutions and how they affect development.

In a later chapter we will assimilate this information with information presented earlier on animal adaptation, as well as material from the next chapter (Climate) to see exactly what types of agricultures have developed to accommodate all these factors.

References

Bhattacharya, P. 1978. Buffalo. Chapter 6 in Williamson, G. and W.J.A. Payne. An Introduction to Animal Husbandry in the Tropics, 3rd. Ed. Longman. London and New York.

Devendra, C. 1978. Goats. Chapter 8 in Williamson, G. and W.J.A. Payne. An Introduction to Animal Husbandry in the Tropics, 3rd Ed. Longman. London and New York.

FAO. 1994 FAO Production Yearbook.

Fernandes-Baca, S. 1978. Llamoids or New World Camelidae. Chapter 10 in Williamson, G. and W.J.A. Payne. An Introduction to Animal Husbandry in the Tropics, 3rd Ed. Longman. London and New York.

Harris, Marvin and Eric B. Ross. 1978 "How Beef Became King" from Psychology Today. Taken from People and Food, p.53, by D. Kirk and E. Eliason. 1982. Boyd and Fraser Publishing Company, San Francisco.

Harris, Marvin. 1990. "Mother Cow" in: Conformity and Conflict, Readings in Cultural Anthropology. 7th ed. James P. Spradley and Dave W. McCurdy eds. Harper Collins Publishers.

Johnson, James E. 1975. Conflict and Complementary in Asian Animal Feeding and Food Production. Prepared for Rockefeller Foundation Review of the Role of Animals in the Future World Food Supply, New York, April 21-22, 1975.

Jordan, T.E., M. Domesh, and L. Rowntree. 1994. The Human Mosaic: A Thematic Introduction to Geography, 6th ed. Harper Collins College Publishers.

Knutson, A.L. 1981. "The Meaning of Food" in People and Food, p.19, by D. Kirk and E. Eliason, 1982. Boyd and Fraser Publishing Company, San Francisco.

McDowell, R.E. 1991. A Partnership for Humans and Animals. Kinnic Publishers. Kinnickinnic Agrisultants, Inc., Raleigh, N.C.

Smith, Houston. 1994. The Illustrated World's Religions. A Guide to our Wisdom and Traditions. Harper, San Francisco.

USDA. 1975. Foreign Agricultural Service. World Agricultural Production and Trade. Statistical Report. May 1975.

Study Questions

1. What are the 3 major religions of the world that have a direct and marked effect on the numbers and utilization of livestock in the countries where they are the predominant religion? What class of livestock does each primarily affect and what is the effect on numbers and utilization of each class of livestock?

2. Although cattle cannot be used for meat in India, the vast numbers of cows found in the country do make several major contributions. For example they are a source of bull calves for use as work animals. What are 2 important reasons why each farm requires at least 2 work animals? In addition to supplying milk, what are 2 other important contributions of cattle to India?

3. Why is cattle manure such good fuel for cook fires?

4. What religion has Kosher food inspection?

5. What is the difference between a market economy and a centrally planned economy? What is meant by the terms First World, Second World and Third World Countries?

6. Which continents are primarily composed of developing countries? What are the only two countries of Asia and the one country of Africa that are classified as developed?

7. What are the 3 levels of agricultural development and describe the characteristics of each from the standpoint of percent of the population that are engaged in agriculture and type of farm power used?

8. Describe the 3 essential services that are provided by financial institutions and why is each important.

9. Describe the 2 essential services provided by marketing institutions and explain why they are important.

10. Discuss the 3 major reasons why industrial development is essential for the existence of a highly developed agriculture in a country.

11. What are the functions of government as it relates to development?

Chapter 4

Climatic Environments of the World

Introduction

In this chapter we consider how climate, a major part of the natural environment, affects agriculture, its animals and its people. The climate has enormous influence on all types of economic activity in which people are engaged. It is absolutely crucial to the selection of agricultural pursuits because climate determines in large measure what plants and animals can and cannot be grown. Thus it has an overwhelming influence on the basic type of agriculture practiced. When we couple a climatic influence with the cultural and economic affectors previously discussed, we will be able to take a good look at the agricultures that have emerged. That we will do in the next chapter. However, first things first. Let's do climate.

People live and produce agricultural animals in a wide variety of climates. These climates impose some degree of climatic stress. It is important to consider the geographical location of the worlds climates and consider briefly some of their more significant effects on agricultural animals serving human societies.

Tropical Climates

Figure 1 presents the locations of the areas of the world that have climates classified as tropical. These are further broken down into the humid and subhumid tropics. Most of the humid tropics are covered by dense rainforests and, at best, are only slightly utilized for agriculture. The subhumid areas are the major portions of the tropics that are utilized for agriculture.

In general, the humid and subhumid areas of the world which lie between the Tropic of Cancer (23°N) and Tropic of Capricorn (23°S) are considered to be the tropics of the world. Obviously, the closer one is to the equator the more tropical the environment, while the areas immediately adjacent to the two Tropics are semitropical.

Isolated areas between the two tropics may not have the high ambient temperatures that are typical of tropical environments, usually because of high elevations. It should also be pointed out that these lines do not exactly set the northern and southern limits of hot environments. One has to but look at the map to see many exceptions. Florida is north of the Tropic of Cancer yet is semitropical. Oklahoma is considerably north, yet has summers that qualify, most years, as hot. However, the true tropics and their associated problems lie between the two Tropics.

In the tropics the major climatic variation from season to season is not in ambient temperature but rather in the amount and distribution of rainfall. This means there are only two seasons -- rainy and dry, with little difference between the two in ambient temperature. These names are very descriptive, during the rainy season it rains nearly every day and during the dry season it does not rain.

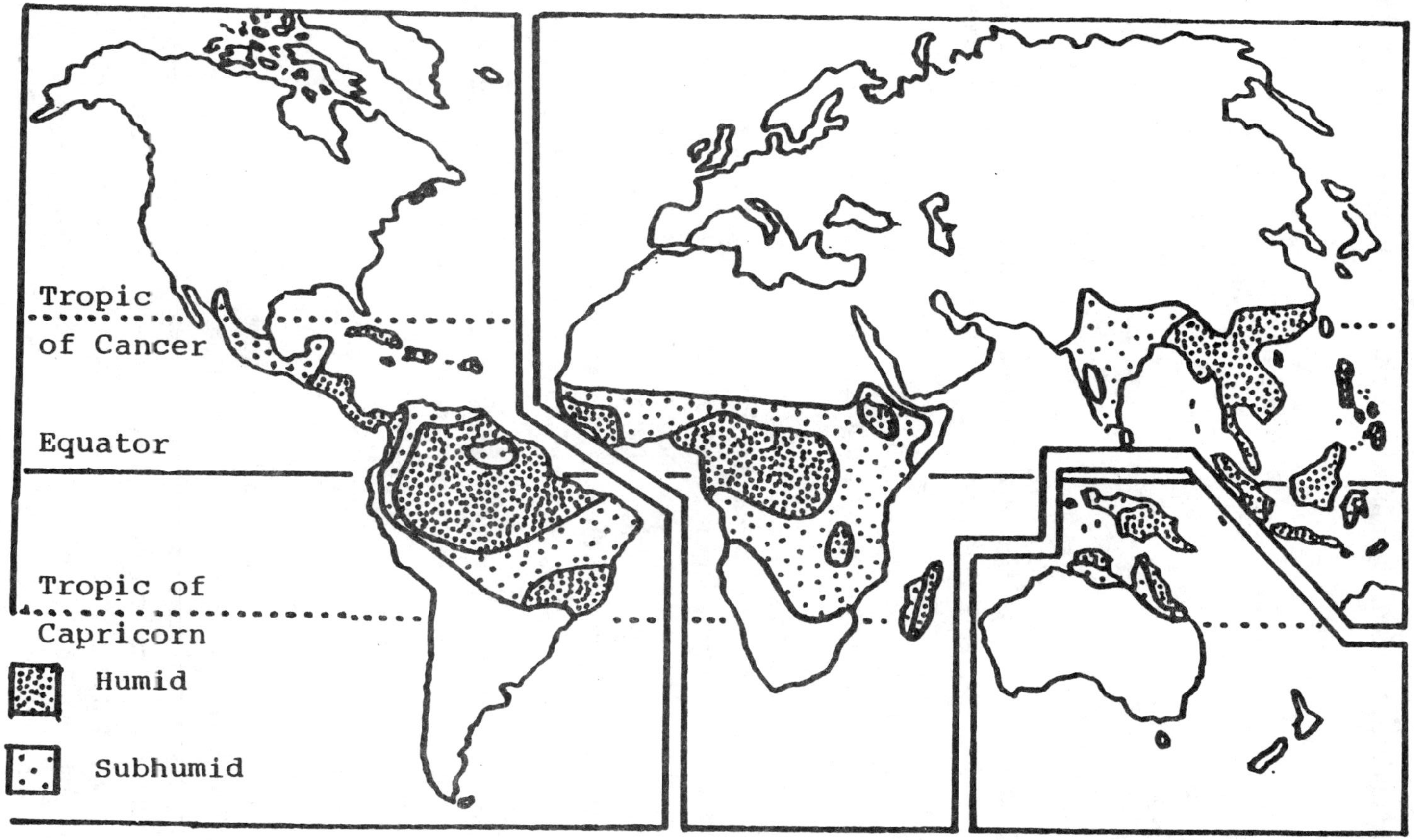

Source: USDA, 1941 Yearbook of Agriculture

Figure 1 Distribution of the Tropical Climates of the World

The following are the classifications of regions between the two tropics based on the length of the rainy season.

Region	Length of Rainy Season
Tropical rainforest	11 months or more
Humid tropics	7 - 11 months
Semi-humid tropics	4 1/2 - 7 months
Semi-arid tropics	2 - 4 1/2 months
Semi-desert to desert	0 - 2 months

The length of the dry portion of the cycle has a very significant effect on the type and amount of vegetation produced, and this largely determines the system of livestock production that is practiced. In many of the developing countries this, in turn, is the major factor determining whether the people live in villages or are nomads. If the dry season lasts long enough that animals run out of feed before the next rainy season, the people are probably nomads.

One of the two most important problems associated with any attempt to produce agricultural animals in the humid tropics is extreme heat stress. There are three primary causes of this heat stress: (1) high ambient temperatures; (2) high relative humidity; and (3) intense solar radiation because the sun is almost directly overhead and there is a minimum of atmospheric attenuation. A fourth contributing factor is there is little wind to cause convective cooling. Generally speaking, the wetter an area the less wind there is. An additional factor is the dense vegetative cover which has a very definite dampening effect on movement of what little wind there is.

The second major problem limiting livestock production in the humid tropics is diseases and parasites. These thrive in a hot humid environment and are a constant threat. They are probably the major factor limiting the introduction of non-adapted animals for the purpose of grading-up local stock.

In the tropics, quantity of food for agricultural animals is seldom a problem, but quality is. The nutritive value of much of the plants is low, largely because of mineral deficiencies in the soil due to excessive leaching by the heavy rainfall. In addition, other climatic factors interact with the rainfall to produce less nutritious forages in the tropics. The Tropical plants have lower nutritive value than temperate ones because they are metabolically adapted to long periods of darkness. Plant nutrients are metabolized at night when none are produced to fulfill the needs of the plant. High temperatures also reduce nutritive value of tropical forages. Digestibility of tropical grasses decreases by about one-half to one unit per degree Celsius increase in temperature.

Deserts

Figure 2 presents the areas of the world that are either desert or very dry. Most of the major deserts of the world are located within two belts roughly 8° on either side of the Tropics of Cancer and Capricorn. The exact location of the belts is between 15°N and 32°N and between

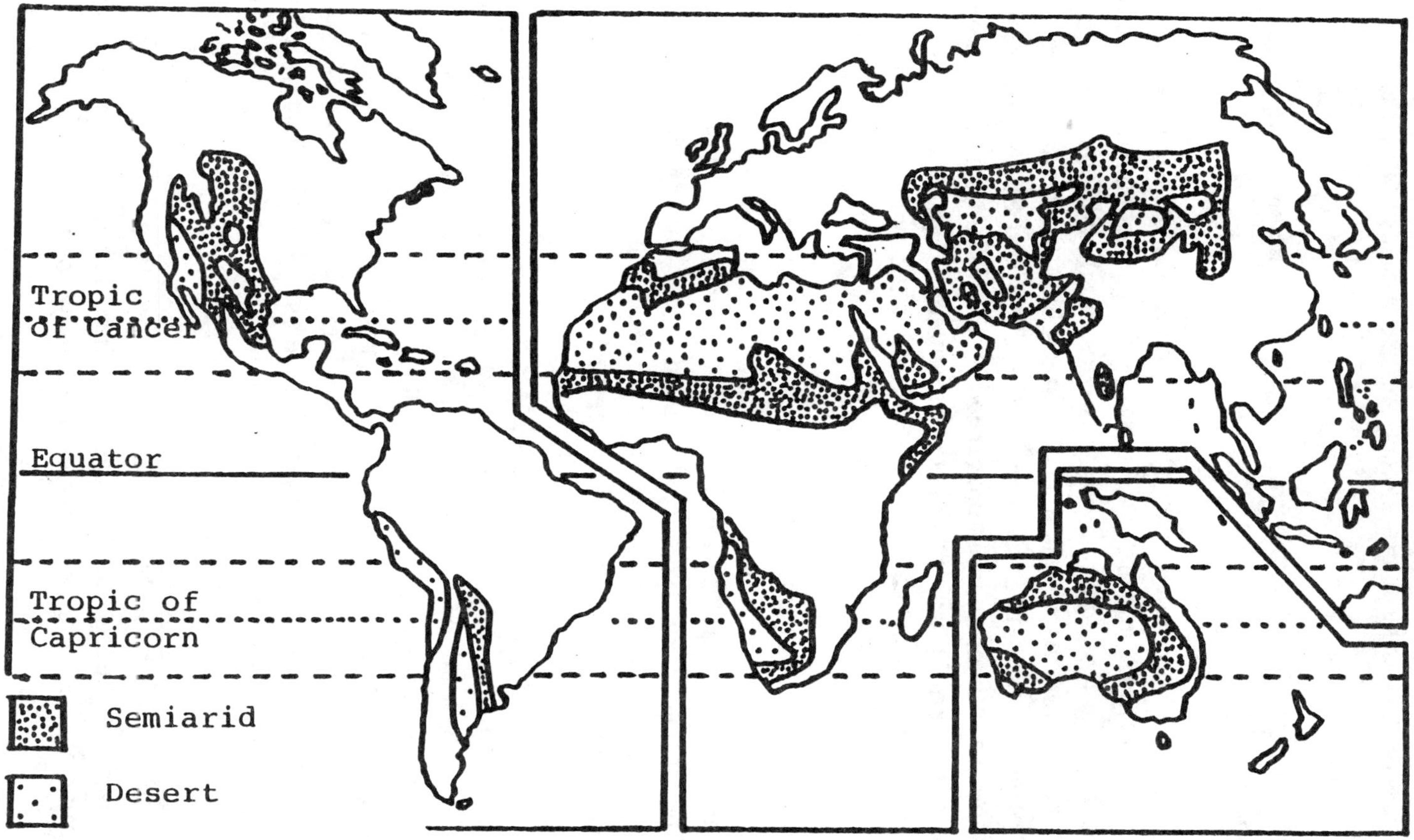

Source: USDA, 1941 Yearbook of Agriculture

Figure 2 Distribution of the Desert and Dry Climates of the World

15°S and 32°S of the equator. The major exceptions are the deserts of the Near East and of Northern China.

There are three types of deserts based on the primary reason for the lack of rainfall. Two of these are illustrated in **Figure 3**.

The deserts in the two bands on either side of the two Tropics result from the pattern of air movement north and south from the equator. Air rises to high elevations at the equator. As it rises it cools and much of the moisture it contains is condensed and falls as rain accounting for the heavy rainfall at the equator. After reaching very high altitudes, the air moves either north or south towards the poles. It descends in the area 15° - 32° N and S. As it descends the air is compressed and warmed so little or no moisture carried by the air can condense into rain. Upon reaching the surface of the earth, the air returns to the equator. It picks up moisture as it moves across large bodies of water so it is heavily laden with moisture when it reaches the equator to repeat the cycle.

The other two types of deserts accounts for most of the deserts of the United States and Northern Asia. Rain Shadow Deserts are found on the leeward side of high mountains. They result from the moisture in the air being condensed as it rises over the mountains. As a result there is little moisture left when the air reaches the lee side. The third type is the Interior Continental Deserts found in areas that are great distances from any large body of water, primarily oceans. Because of this isolation, little, or no, moisture laden air ever reaches them.

There are two major problems that limit livestock production in the semiarid regions of the world. To a lesser extent they are also the problems of the semihumid tropics. These are: (1) seasonal and limited rainfall, resulting in a shortage of drinking water and water to grow food for the animals; and (2) extreme heat stress. The reasons for the extreme heat stress are primarily the high ambient temperatures and intense solar radiations associated with close proximity to the equator. The solar radiations are more intense than in a humid area in the same latitude because of the lack of water vapor in the air to dampen radiations. Another feature of the ambient temperature in the drier areas is the extreme daily variation in temperature. This ranges from very hot during the day to cool to cold at night.

The seasonal and limited rainfall contributes to two major problems. There is a scarcity of both food and water, particularly during the dry season. There are also very few trees or other vegetative cover to serve as shade to protect against solar radiations.

The semiarid tropics are also bothered by parasites and insects, but not to the extent of the humid tropics. This problem is also of less magnitude in the subhumid tropics. This is one of the important reasons why a high percentage of the livestock in the tropics are found in the subhumid and semiarid regions.

Another problem in the drier areas of the world is high winds because of the association between the wetness of an area and amount of wind. These winds are a major cause of erosion in very dry, overgrazed areas and a contributing factor to the spread of deserts (desertification).

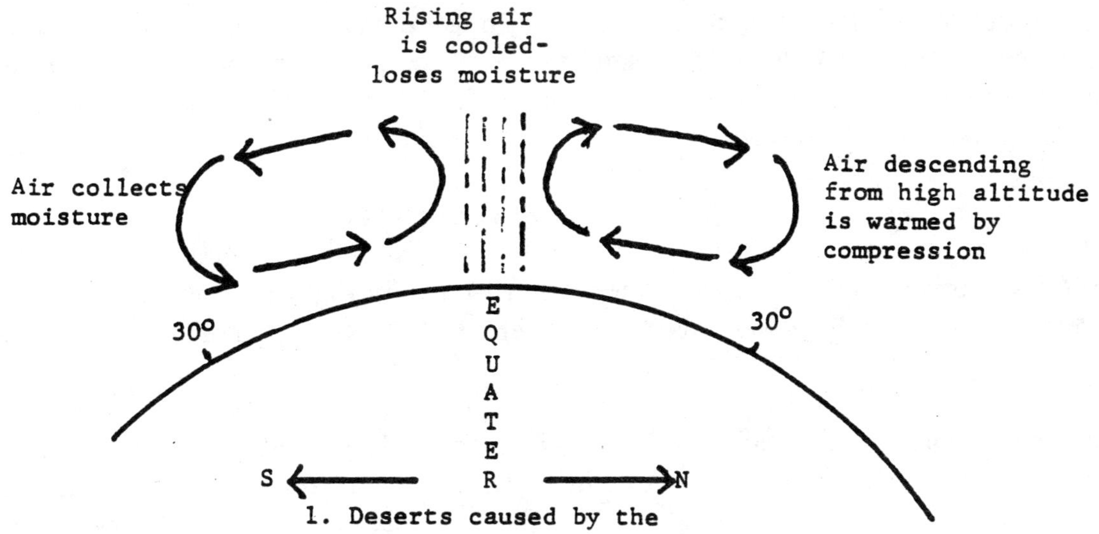

1. Deserts caused by the pattern of air movement north and south of the equator.

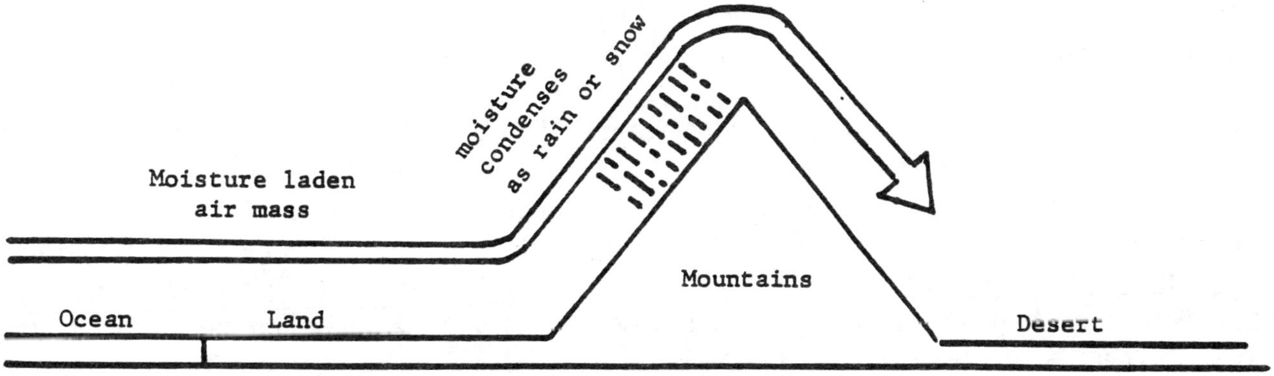

2. Rain Shadow Deserts - On the leeward side of mountains.

Figure 3. Causes of Two Types of Deserts

The problems of the semiarid regions are of even greater magnitude in the desert areas. However, they are of little practical significance since few, or no, livestock are produced in these areas. The problem the desert poses to a country is it represents land that is completely wasted as far as agriculture is concerned.

Cold Environments

The distribution of cold climates of the world is shown in **Figure 4**. Generally, they are in the far northern regions of the world. The major exception is the very high mountains of central Asia.

The major problems associated with livestock production in these areas of the world are: (1) cold stress; and (2) availability of food. Within limits, it is usually easier for an animal to adapt to and produce in cold temperatures than in hot. However, from a practical standpoint, the very cold areas are so devoid of food that whether animals can adapt to the cold is an academic question. The reason for the lack of food is associated with both a general scarcity of plant life and prolonged snow and ice cover. Therefore, the very cold areas of the world are largely devoid of livestock.

Temperate Climates

To complete the world climatic picture, the areas of the world with temperate climates are shown in **Figure 5**. Generally, these are the most productive areas of the world. All developed countries have a temperate climate in at least part of their boundary. This is the climate found in most of the continental US except the arid and semi-arid West and the semi-tropical portions of Florida. Livestock producers are faced with many challenges in these zones. However, generally speaking, the agricultural problems of the rest of the world are compared to temperate conditions because they provide the best available environments.

Climate and Natural Vegetation

Climate has a major indirect effect on livestock through its influence on the vegetation of a region. The relationship between rainfall and ambient temperature of a region and the natural vegetation of the region is summarized in **Figure 6**. (Natural vegetation is the vegetation that was there before human societies modified it.) If the land was suitable for cultivation, the natural vegetation has largely been destroyed in the process of clearing the land for farming.

In **Figure 6**, the horizontal axis is divided into blocks according to the amount of rainfall, with classifications from arid to superhumid. The vertical axis is divided into blocks based on ambient temperature moving from very hot to very cold as you move from top to bottom. Within each block is listed the natural vegetation for that combination of rainfall and ambient temperature.

As shown in **Figure 6**, the natural vegetation of the wet areas (humid or superhumid) of the world that are either hot or have a temperate climate is some type of forest. The specific type of

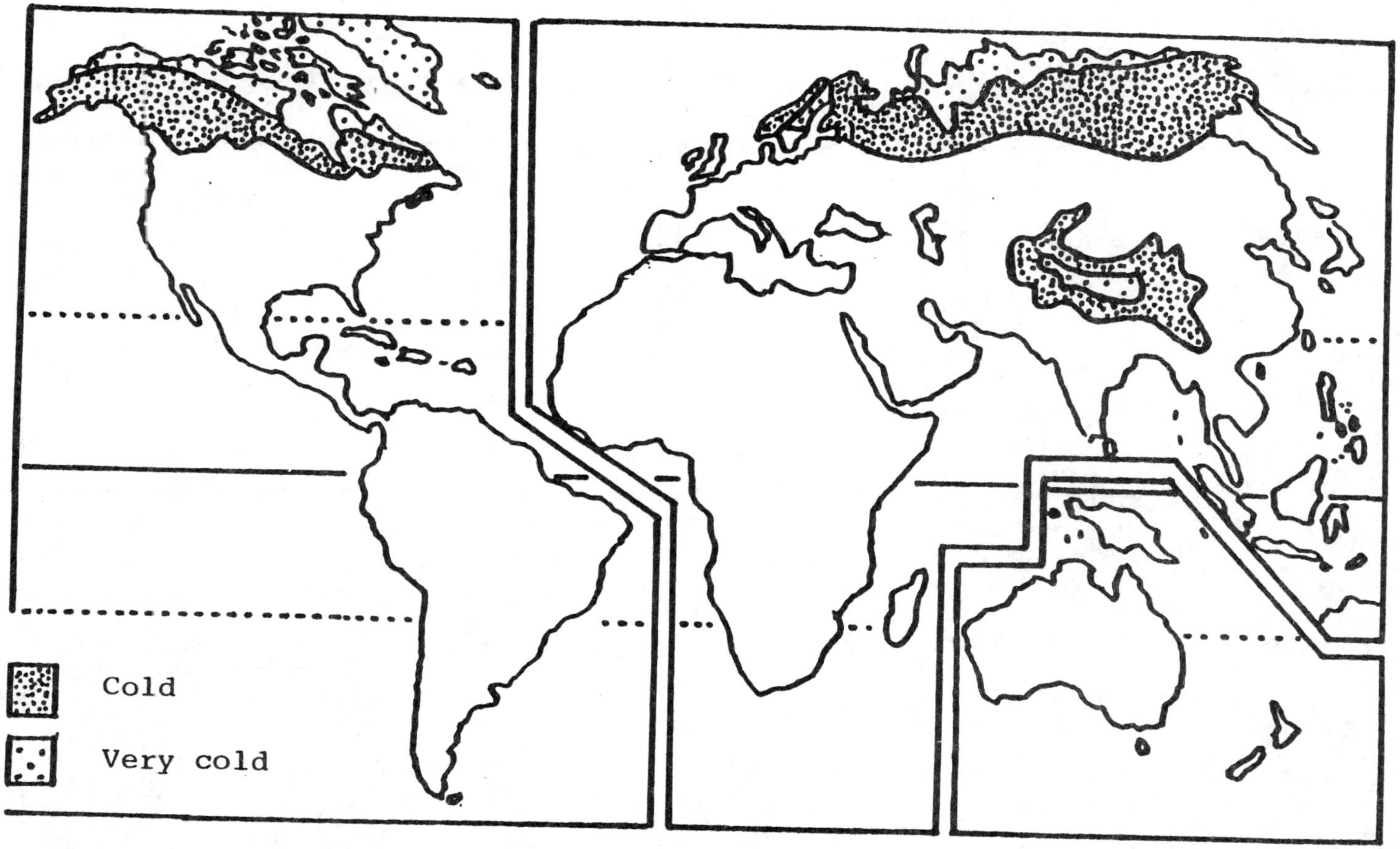

Source: USDA, 1941 Yearbook of Agriculture

Figure 4. Distribution of the Cold Climates of the World

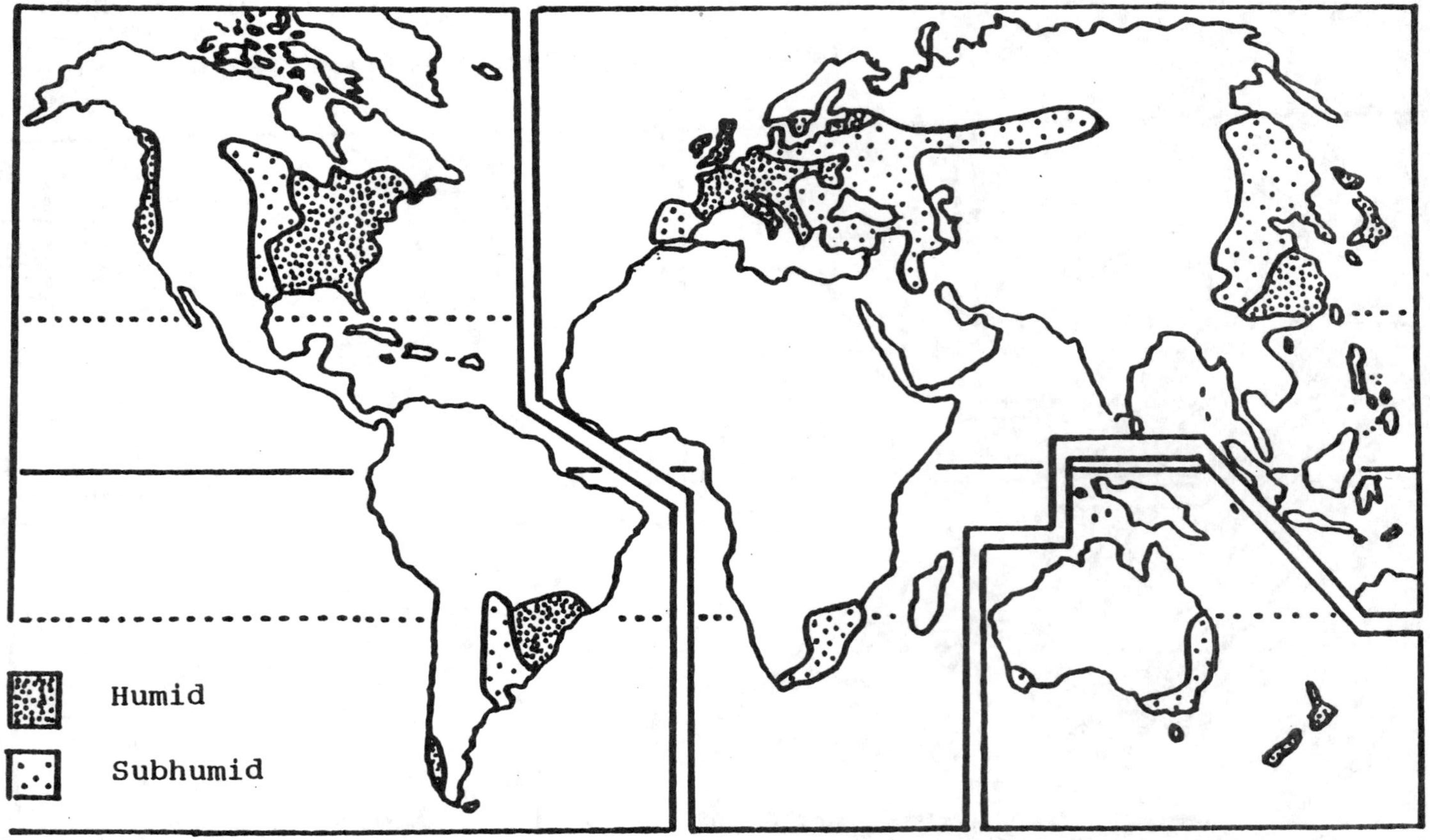

Source: USDA, 1941 Yearbook of Agriculture

Figure 5. Distribution of the Temperate Climates of the World

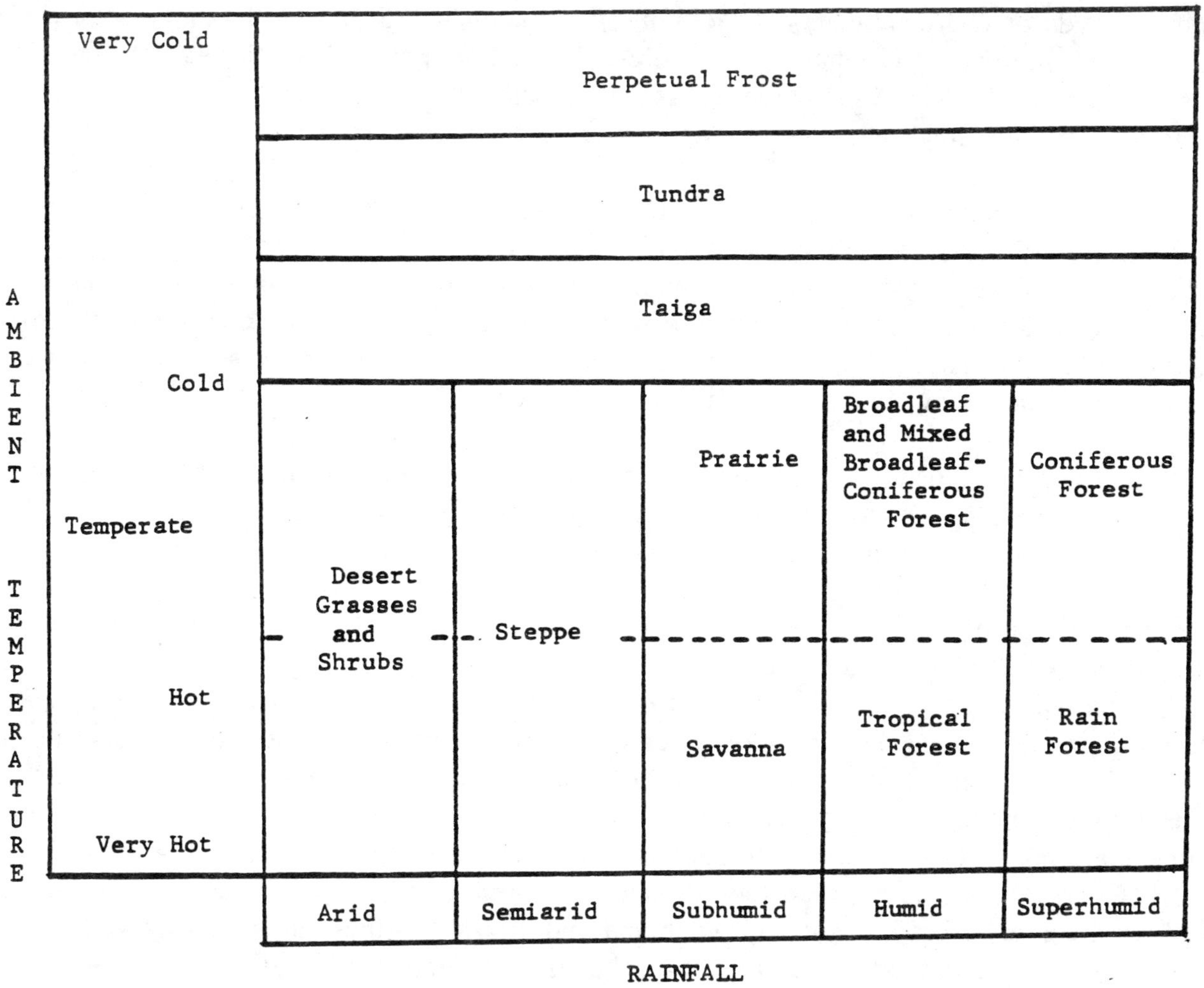

Figure 6. The Relationship Between the Rainfall and Ambient Temperature of a Region and the Natural Vegetation of that Region. (Source: USDA, 1941 Yearbook of Agriculture).

forest depends on the combination of rainfall and ambient temperature. In the wettest areas (superhumid) the natural vegetation is rainforests in the hot areas and coniferous forests in the temperate areas. If the area is classified as humid, the natural vegetation in the hot areas is tropical forests and broadleaf or mixed broadleaf-coniferous forests in the temperate regions.

The specific distribution of these four types of natural vegetation is shown in **Figure** 7. Rainforests are difficult to clear and farm, and seldom are. The remainder can be cleared. Most of the forests of the humid, temperate areas have been cleared and constitute the farmlands of eastern U.S. and of Europe. One other type of forest is taiga, which are the coniferous forests, primarily of spruce and fir, found in cold areas regardless of rainfall. The vast evergreen forests of extreme northern United States and Canada are this type.

As rainfall decreases, natural vegetation becomes some type of grassland. First the tall grasses, then, with decreasing rainfall, the short grasses and finally desert grasses and shrubs. The distribution of these grasslands is shown in **Figure 8.**

The specific type of tall grass natural vegetation depends on the ambient temperature. In hot areas of the world it is called savanna. In temperate areas it is called prairie. Where there is enough rainfall to support farming, most of the prairie type natural vegetation has been plowed and put into cultivation. This comprises some of the most productive agricultural land in the world. The corn belt of the United States and the pampas of Argentina are prime examples. Likewise the most productive grazing areas of the world are the prairie areas where the rainfall amounts are marginal for cultivation (subhumid to semiarid). In the United States this is the central Great Plains area between the 30" and 20" rainfall lines. It includes the great cattle ranching areas of Nebraska, Kansas, Oklahoma and Texas. In these areas, the rainfall is too unpredictable to permit extensive cultivation, but is sufficient in most years to give excellent growth of grass. Parts of these areas may be plowed during years of high wheat prices and are returned to grass during years of low prices.

As rainfall decreases, the grass species change from the tall grasses to the short grasses. Regardless of the temperature, this is called steppe vegetation, and these areas account for most of the land area of the world that is devoted to range livestock production. This is primarily because these areas are too dry to be farmed and can only raise grass.

The areas covered with desert grasses and shrubs support only a limited number of livestock as a result of the shortage of both drinking water and forage. Except for Europe, all continents have a sizable amount of this type of vegetation.

Tundra is the lichen, moss vegetation that grows in the limited area between perpetual frost and taiga. This is the vegetation of the far north, and may also be found at high elevations just above the tree line in the Rocky Mountains. It is a very fragile vegetation. If damaged it may require many years to recover. One of the objections environmentalists had to building the Alaskan pipeline was possible damage to the tundra. Other than providing feed for a few reindeer, the tundra makes no contribution to agriculture.

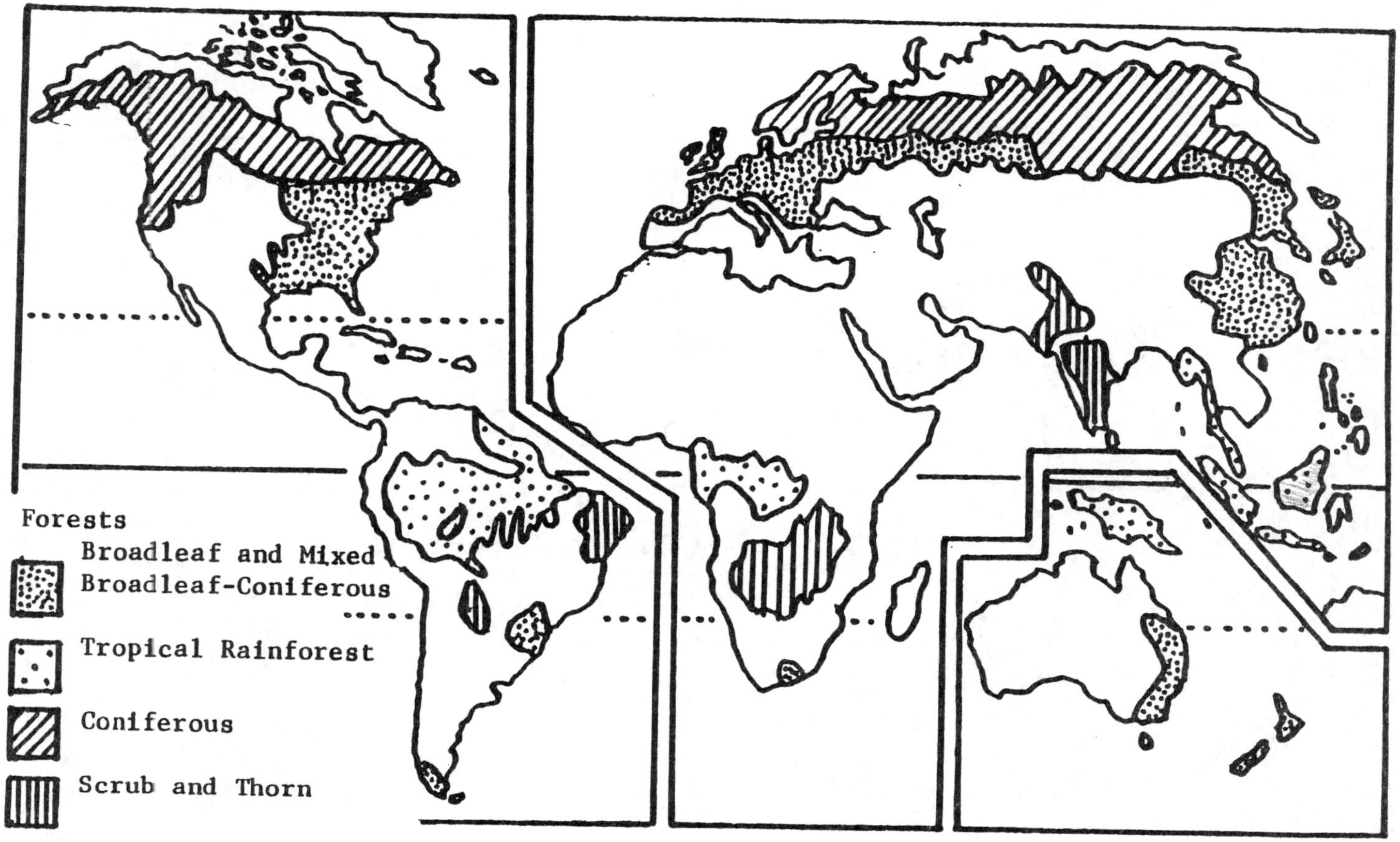

Source: USDA, 1941 Yearbook of Agriculture

Figure 7. Areas of the World in which the Natural Vegetation was some Type of Forest.

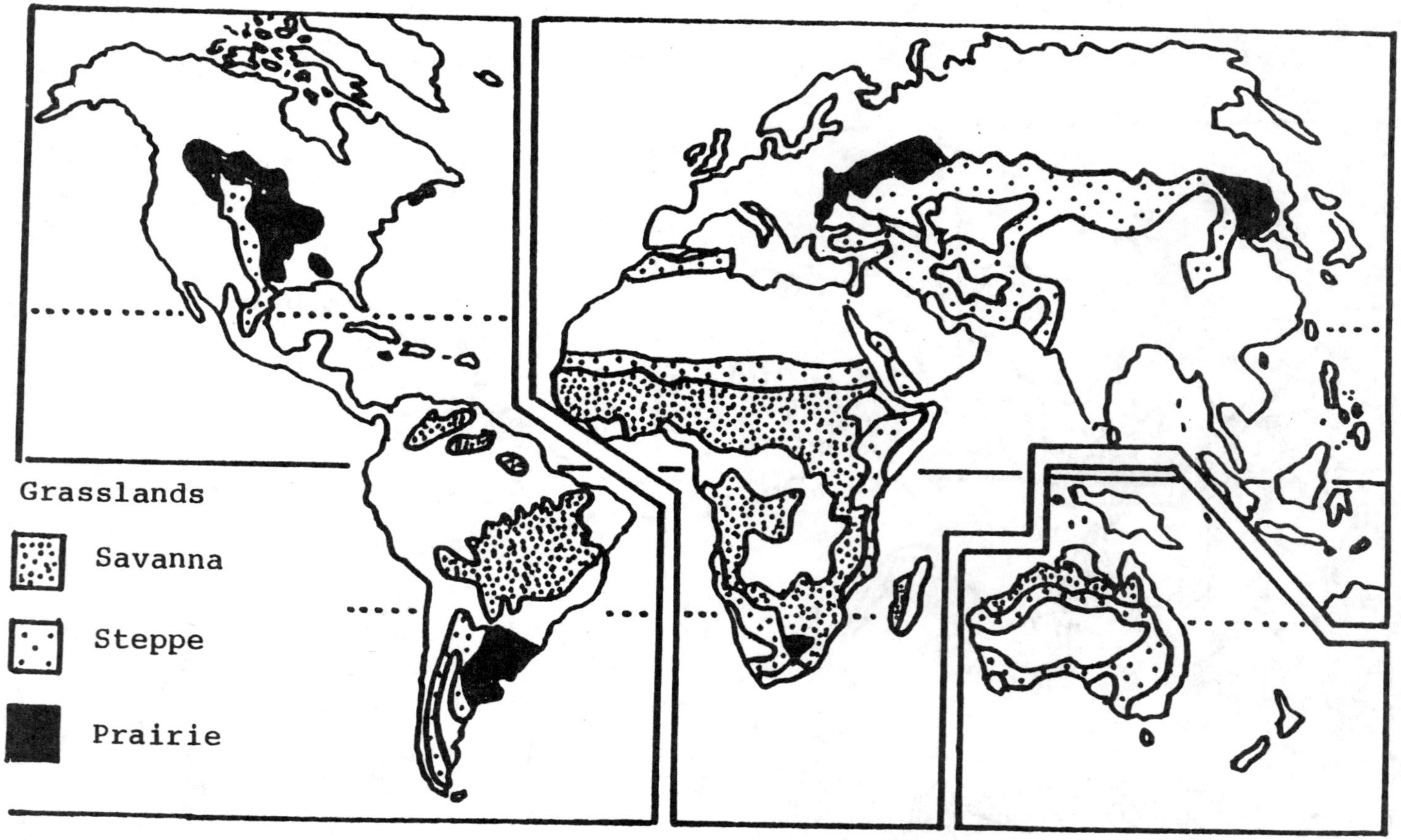

Source: USDA, 1941 Yearbook of Agriculture

Figure 8. Areas of the World in Which the Natural Vegetation was Grassland.

Desertification

Desertification is not a new term, however, it has been only in recent years that it has been the subject of concern on an international level. It is likely that one of the major reasons for this increased concern was the disastrous droughts from 1968 through 1973 in the Sahel zone of Africa just south of the Sahara Desert. Pictures of starving people were continually shown on TV in developing countries which heightened awareness of the general public. The term, desertification, suggests shifting desert sand dunes taking over areas that were once green fields and pastures. This definition is far too restricted. As now used, desertification is defined as all of the ecological changes that reduce the ability of arid and semiarid lands to sustain agriculture and support human habitation.

There is more than one-third of the earth's land area that is arid or semiarid and is occupied by approximately 700 million people. Nearly one half of this land is so dry and barren that it cannot support human life. It is probable that some 80 million people live on lands that, because of desertification, are already almost useless for agriculture. Thus, it is imperative that we learn how to better utilize very dry areas and prevent and reverse the process of desertification. Unfortunately, solutions include hard, expensive choices and the need for great political will. Both are hard to come by.

The problem of desertification is two-fold: (1) actual reversion to desert of land that was formerly productive to some extent; and (2) the degradation of cropland and grassland. The latter problem is probably the most serious because it is so much more widespread than is actual reversion to desert. It involves a marked reduction in productivity and carrying capacity as soil nutrients are dissipated and topsoil lost through erosion, or as quality of rangelands forages declines as the result of undesirable plant species replacing the more desirable. One acre which reverts to desert may cause the degradation of 100 or more adjacent acres.

There are two major causes of desertification: (1) prolonged periods of drought; and (2) improper land use practices. These improper land use practices are almost always the result of expanding human and animal populations and include deforestation, overgrazing, burning and improper farming practices. One of the most common improper farming practices is attempting to farm land that should be left in grass.

How serious is the problem? The spread of the desert in Africa can be fairly precisely measured, but the deterioration in productivity can only be estimated. The Sahara desert has gradually shifted southward and taken over an additional 250,000 square miles (166 million acres) in the last 50 years.

The desertification process occurs in a stepwise fashion. Specific vegetation zones gradually shift southward as a result of overgrazing, woodcutting and accelerated soil erosion. The process is in a definite stepwise fashion--desert creeps in and replaces steppe which in turn creeps in and gradually replaces savanna and savanna creeps into the forest and replaces it. Eventually, if not checked, land in Africa that is now occupied by trees or the tall grasses of the savanna could become desert.

Desertification is not just a problem immediately adjacent to the Sahara desert. It is a major problem in parts of southern Africa, particularly Botswana, and in Kenya, Tanzinia and Ethiopia. It is also a constant threat in the Middle East and western Asia. Nor is the problem confined strictly to the less developed countries of the world. Australia and the United States have their share of improperly managed rangelands. The Dust Bowl of the 1930's in the Great Plains of the United States was a vivid example of the folly of plowing lands best left in grass. There are still millions of semiarid acres being cultivated that would be better off in grass. Although the Dust Bowl occurred many years ago the underlying causes still remain a threat in much of our arid and semiarid southwest and west.

There are many problems facing governments in their efforts to halt the spread of deserts and the deterioration of semiarid regions. One of the major ones is that of preventing overgrazing. In many areas of the world the necessary reforms are contrary to customs and traditions. Many of the severely overgrazed areas are utilized by nomads, and have been for generations. They move to where grass is available, and usually pay no attention to national boundaries or national governments. If one group does not completely denude an area of grass another group will. They feel that the grass was placed there to be eaten and governments have no right to dictate how much or by whom.

Even if the people are not nomads, it is difficult to legislate proper range management. In the developed countries we have private ownership of most of the land and any government intervention would be bitterly resisted. In the developing countries, much of the land is common land so the government has some claim to the right to control its use. However, it is difficult to do so because it is contrary to tradition. A man's wealth is measured in terms of the size of his herds and flocks. As the human population grows so does the livestock population, and not on a one to one basis since it takes more than one animal to support a person. One of the essentials in preventing desertification is a great reduction in the size of herds and flocks utilizing susceptible lands.

A good summary of the problem of desertification is given in an article, The Spreading Desert, published in the AID publication, "War on Hunger." (Eckholm and Brown, 1977).

> "The negative environmental trends called desertification are widespread, long standing and, in many areas, accelerating. The technologies needed for reversing them are for the most part already available and well summarized in the 'Plan of Action' adopted by the UN Conference in Nairobi. But too commonly lacking is a political commitment to their reversal commensurate with the size of the challenge. Faced with immediate crisis---famine, strikes and political intrigues--governments find it difficult to devote substantial resources to combating a seemingly long-term and nearly invisible problem like ecological deterioration. They are especially reticent to do so when a shift in national priorities and investment patterns goes against the short term personal interests of powerful elites. Governments that procrastinate too long, however, may one day be forced by events to see that their deteriorating agricultural landscapes are mirrored in deteriorating social and

economic conditions. The varied consequences of desertification--undernutrition and famine, unemployment and migration, deepening poverty and human desperation--are neither distant nor invisible."

Conclusion

Climate has an important influence on agriculture by determining which plants and animals will be available to the agriculturists of a given climate. Clearly, some climates are more conducive to agriculture, especially animal agriculture, than others are. How well man is able to use them often depends on his overall economic well-being because this determines the extent to which he can either modify the environment or protect the animals from it.

3903/chap4B.doc.sum96

References

Eckholm, Erik and Leston R. Brown. 1977. The Spreading Desert. War on Hunger, Agency for International Development, August and September-October, 1977, Vol. 12:No. 8 and 9.

USDA. 1941. Climate and Man. 1941 Yearbook of Agriculture.

STUDY QUESTIONS

1. What is the specific name of the lines on maps that marks the northern and southern boundaries of the portion of the world within which the tropics and semitropics are found? In this region what is the major climatic variation from season to season?
2. What are the 2 most important problems associated with producing agricultural animals in the humid tropics? What are the 3 primary causes of the extreme heat stress of the humid tropics and why is there so little wind?
3. In what 2 bands relative to the Tropics of Cancer and Capricorn are most of the major deserts of the world located? Explain clearly and completely the specific reason for the lack of rainfall in these two bands that causes these deserts, accurately describing the patterns of air movement north and south of the equator that are involved. What are the other 2 types of deserts and what causes them?
4. What are the 2 major problems limiting livestock production in the semiarid regions of the world?
5. Describe the relationship between the ambient temperature and rainfall of a region and the type of natural vegetation of that area. When is the natural vegetation some type of forest? some type of grassland?
6. What combination of ambient temperature and rainfall results in each of the following: tropical rainforest; tropical forest; coniferous forest; broadleaf and mixed broadleaf forest; savanna; prairie; steppe; taiga; and tundra? Be able to describe each of these types of natural vegetation.
7. Why is there very little wind in the tropics, but a great deal in the arid and semiarid regions of the world?
8. Describe the following types of natural vegetation and the climatic conditions that produce them: tropical rainforest, broadleaf and mixed broadleaf-coniferous forest, savanna, prairie, steppe, taiga, and tundra. Which kinds occur in the United States and where?
9. Desertification implies the conversion of useful land to desert. What is a more accurate and complete definition of desertification? Describe the 2 major causes of desertification.
10. Why do governments have so much trouble preventing desertification?

Chapter 5

World Agricultural Types and Systems of Livestock Production

Introduction

In this chapter we will bring to culmination the discussion of animal adaptation begun in Chapter 2. In that chapter we discussed how animals adapt. In Chapters 3 and 4 we discussed major factors to which animals must adapt to be a successful part of agriculture. In this chapter, we will see the major agricultural types of the world that have evolved in response to man's needs and desires, coupled with the animal's abilities to adapt to cultural, economic and environmental (especially climate) stressors. We will also discuss how animals fit into those agricultures.

Agricultural Systems - An Overview

The predominant system of livestock production found in a given area is primarily determined by the principal agricultural system of that area. In turn, the agricultural system is determined by a number of factors including climate, socioeconomic factors (primarily level of economic development, political systems and religions), topography, and soil type. While all are important, climate, particularly rainfall and temperature, and level of economic development are most important. Climate determines to a large degree what can be grown and therefore the basic type of agriculture practiced but the level of economic (and agricultural) development determines the specific system.

There are a number of classifications of agricultural systems, with one of the most widely accepted being that of Whittlesey (1936). He classified the agricultural systems of the world into 13 types. **Table 1** gives a listing of the systems proposed by Whittlesey modified by grouping them according to the level of agricultural development in which they are found. It has been further modified to reflect changes in agriculture since 1936. The relationship of these systems to amount of rainfall is summarized in **Table 2.**

Of all these systems, only two, nomadic herding and ranching, are concerned almost exclusively with livestock. Their worldwide distribution is shown in **Figure 1**. They have one other thing in common in that they are carried out in the areas of the world where cultivation is either not possible or not profitable. These are primarily regions that cannot be cultivated because they are too dry. As shown in **Table 2**, ranching or nomadic herding is carried out in areas with less than 25 inches of rainfall. When rainfall amounts are between 20 and 25 inches there may be some farming (grain production or subsistence crop and livestock production depending on whether it is in a developed or subsistence agriculture). However, when rainfall amounts are less than 20 inches the land is usually not cultivated.

Although the most common reason why an area will be devoted to grazing livestock and not cultivated is it is too dry, there are three other reasons why such will occur. These are areas that are: (1) too rough to cultivate; (2) located at high elevations where the growing season is too

Table 1. Agricultural Systems of the World

Agricultural Systems Found in Area With Primitive Agriculture. Their goal is subsistence but usually agriculture is supplemented with non-agricultural food sources (hunting, gathering, stealing).

1. Nomadic Herding
2. Shifting Cultivation
3. Rudimental Sedentary Tillage

Agricultural Systems Found in Areas with Subsistence Agriculture.

1. Subsistence Crop and Livestock Farming
2. Intensive Subsistence Tillage - Rice Dominant
3. Intensive Subsistence Tillage - Without Paddy Rice (Peasant Grain and Livestock Farming)

Agricultural Systems Found in Areas with a Developed Agriculture. The goal is to make money with the agricultures.

1. Ranching
2. Commercial Grain Farming
3. Commercial Livestock and Crop Farming
   a) Commercial Livestock Finishing
   b) Commercial Dairy Farming
4. Commercial Plantation
5. Specialized Horticulture
6. Mediterranean Agriculture
7. Market Gardening

Table 2. Influence of Amount of Rainfall on Agricultural Systems in each of the Levels of Agricultural Development

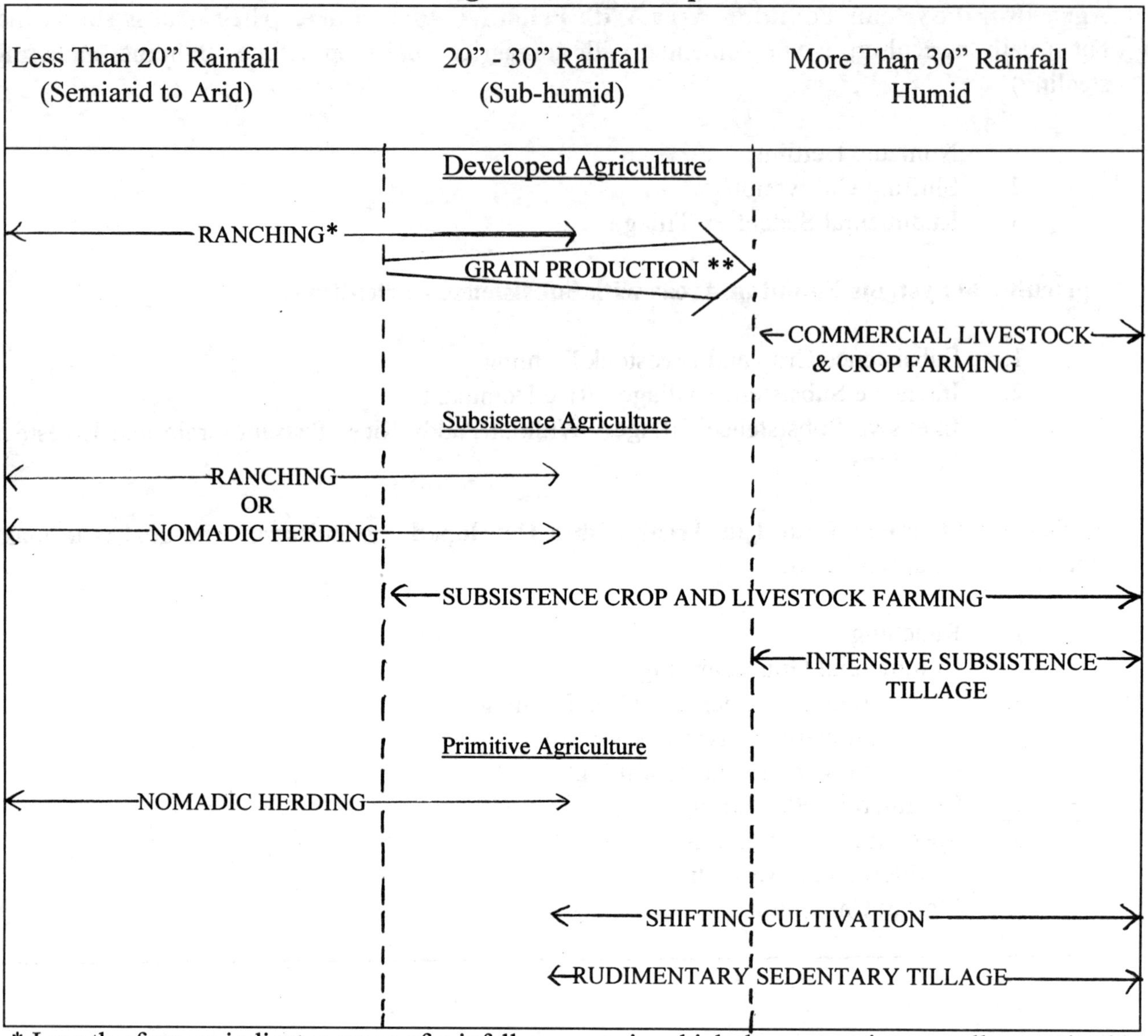

* Length of arrow indicates range of rainfall amounts in which the system is generally practiced.

** Width of arrow indicates extent to which the system is practiced in an area with a given amount of rainfall.

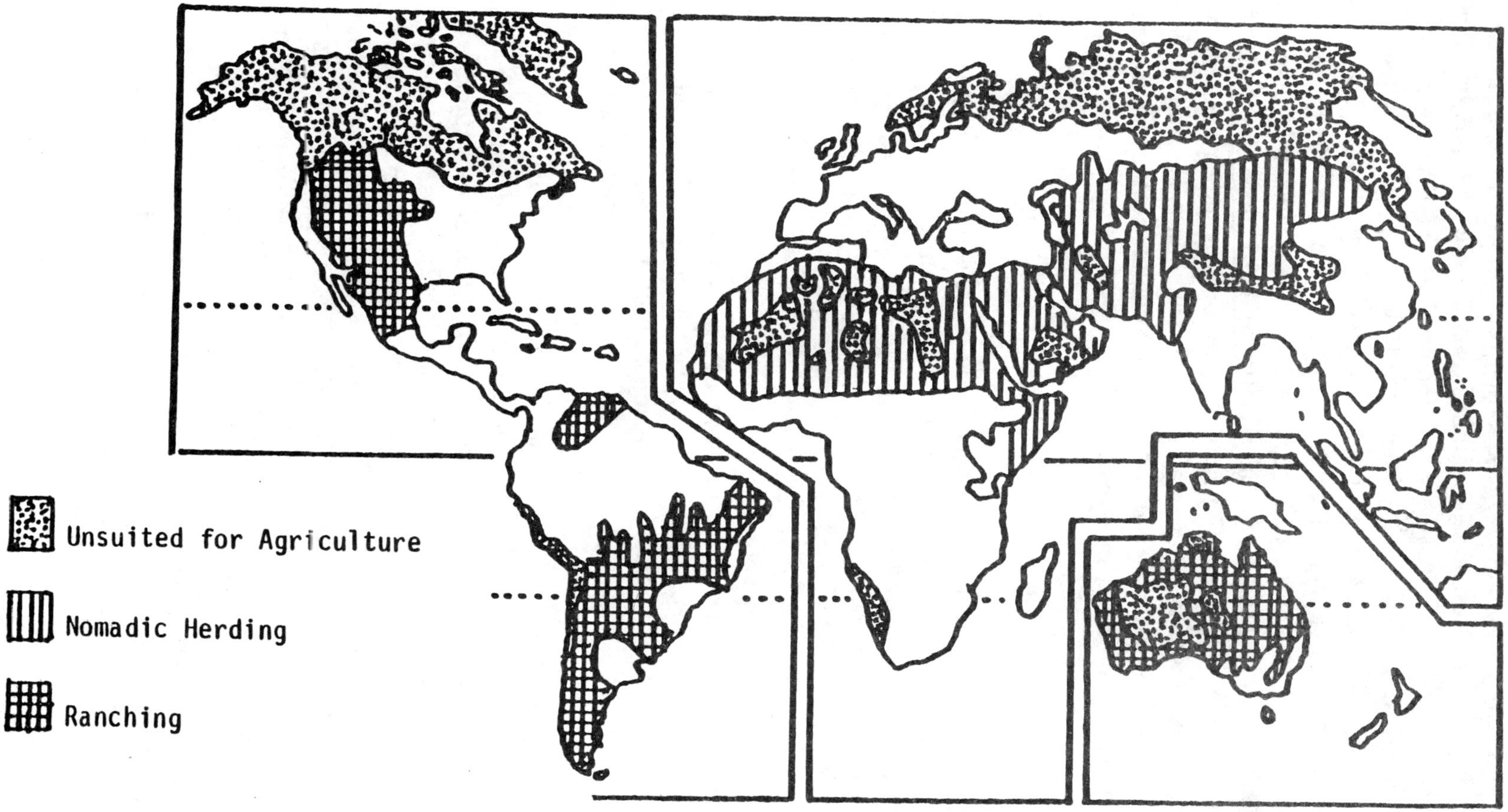

Figure 1. Worldwide Distribution of the Two Agricultural Systems Primarily Devoted to Grazing of Livestock (Adapted from Whittlesey, 1936).

short; and (3) areas so far removed from good markets that the cost of shipping the products to market would be unprofitable. In the latter case, grazing is usually only a temporary utilization of the land and will be discontinued once transportation facilities are built. For example, for a number of years following the Civil War, much of the Corn Belt region of the United States was devoted to grazing of livestock. The reason was not that they did not realize that they could grow corn, but there was no economical way to ship vast amounts of corn to the Eastern seaboard where there was a good market. Once the railroads were extended into the Midwest, the land was plowed up and planted in corn. Much of the area of Brazil that is now utilized for ranching is so utilized for the same reason. Commercial grazing of livestock can never compete with crop production as the most profitable way to utilize land if crop production is possible. Many of the shifts of agriculture that have occurred in Argentina in the last quarter century have been to crop production and have occurred for precisely this reason.

Primitive Agriculture

The worldwide distribution of the agricultural systems found in areas with a primitive agriculture are shown in **Figure 2**. The areas that are dry are utilized by Nomadic Herding. If the land receives enough rainfall to support cultivation, Shifting Cultivation and Rudimentary Sedentary Tillage are practiced. Almost all the agriculture of Africa is one of the primitive systems as is half of Asia and South America.

Nomadic Herding is the primitive utilization of the dry and/or cold regions of the world. It is found especially in the deserts, steppes and savannas of Africa, Arabia and the interior of Eurasia. It is also found in the very cold areas north of the tree line in the countries that were formally part of the Soviet Union. It is practiced by nomadic people who move their families with their herds and flocks of cattle, sheep, goats, camels, reindeer, yaks or horses to where feed and water is available. The herd may consist entirely of a single species or a mixture of these species. Many that place a high value on the horse are those that have a warlike past. Those of sub-Saharan Africa are the only ones who depend almost exclusively on cattle. The reindeer is used only by those that inhabit the Tundras of Eurasia. The goal is subsistence rather than realizing a profit from the sale of their livestock. The people live in tents or snow huts that are easy to transport or to abandon and replace. They own no land although certain tribes may claim certain areas as their lands and resist trespassing. This sometimes creates serious problems when arbitrary political boundaries are constructed that divide traditional grazing areas into two different countries. Seldom, if ever, is any type of farming practiced. Today, Nomadism is a dwindling and probably dying way of life.

Shifting Cultivation consists of clearing small plots of land, farming them for a few years until the fertility declines, then abandoning that plot and clearing another. This is often called slash and burn agriculture because the forest is cut and then burned as a means of clearing. The ash provides nutrients for the crops. These people may be forced to move to a new spot or, if land is available nearby, remain in the same place. Eventually, the abandoned plots grow up in brush, trees and other plants that help to renew soil fertility, and such renewed plots are later cleared and the cycle repeated. In its best form, this is a good system of land rotation. The land should have 10-20 years to rejuvenate. The goal of this agriculture is subsistence. However, many

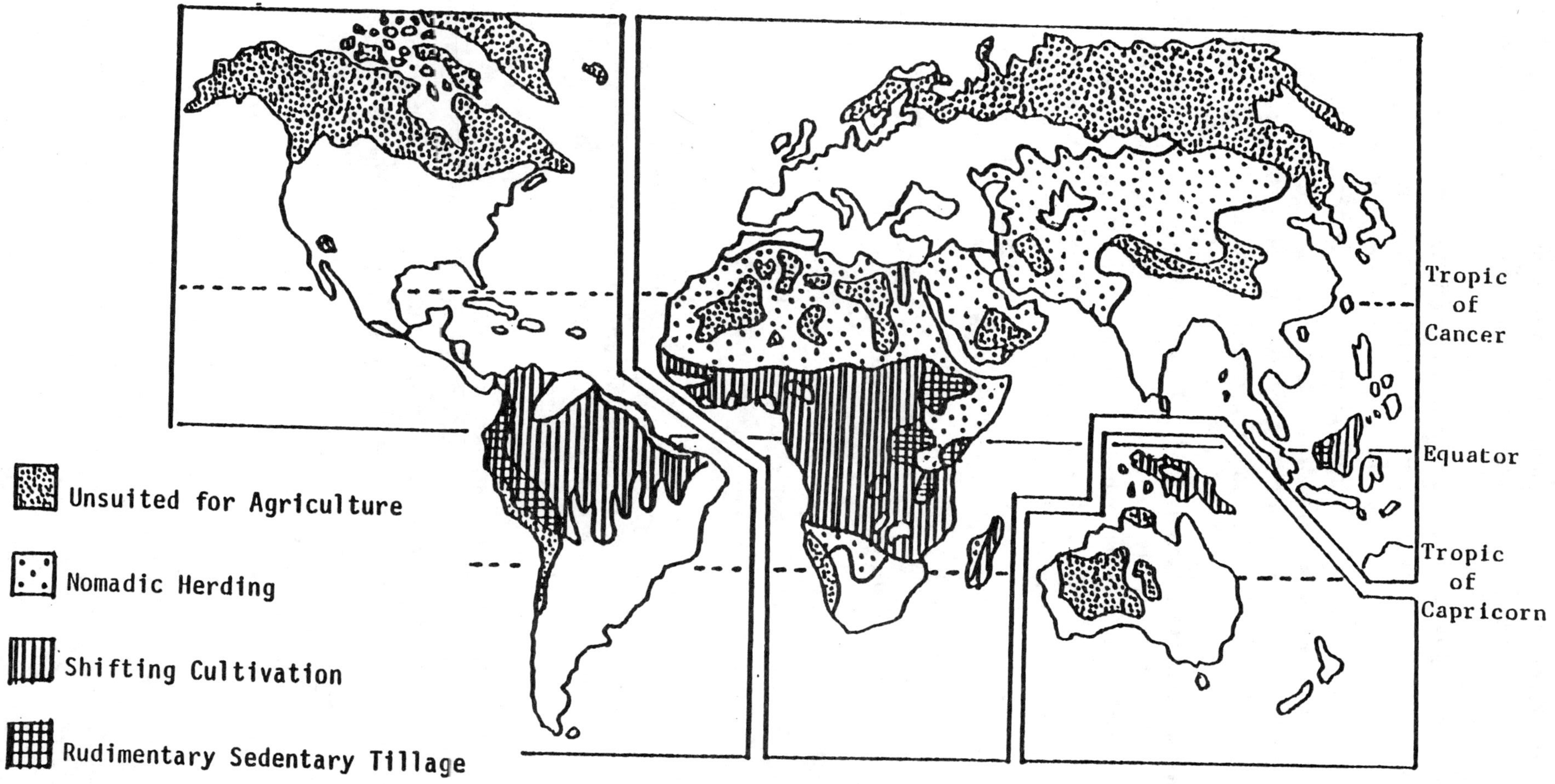

Figure 2. Worldwide Distribution of Agricultural Systems Found in Areas with a Primitive Agriculture (adapted from Whittlesey, 1936).

foodstuffs are generally gathered from the surrounding forest to supplement the agriculture. This is found in the remote tropical areas of South America, Africa, Southeast Asia and Indonesia.

Rudimentary Sedentary Tillage differs from Shifting Cultivation in that the farmers remain in one place and do not shift their cultivation from plot to plot. This system requires land that is basically more fertile. However, the farming methods and crops produced are usually similar to that of Shifting Cultivation and are very primitive involving mainly hand labor and very crude tools. Both of these systems are very efficient in terms of energy efficiency, returning more calories in the form of food for the calories spent on cultivation than does mechanized agriculture. They are also sustainable if the population doesn't increase.

Animals in Rudimentary Sedentary Tillage and Shifting Cultivation. Farmers in both of the above types of primitive farming usually also own some animals though the role of animals is not a major one. Generally, all animals owned by the people of a village graze together on common grazing grounds near the village. One of the jobs assigned to the young boys and old men of the village is herding the animals during the day when they are grazing. Usually, they are brought into the village compound at night for protection from predators - human as well as wild animals. In addition, milk and manure can be collected. Most meat comes from fishing and hunting.

Subsistence Agriculture

Figure 3 presents the distribution of the agricultural systems found in areas with a subsistence agriculture. The two major types are Intensive Subsistence Tillage and Subsistence Crop and Livestock Farming.

Intensive Subsistence Tillage both that involving rice production and that without rice production, is the type of agriculture practiced in many of the Third World Countries with a vast human population. As can be seen in **Figure 3**, this agricultural system is concentrated in south and east Asia. Large acreages are found in India, China, Japan and the Philippines. Rice is grown if the necessary water is available because it will out yield any other grain. With or without rice, this system is very labor intensive, both for human and animal labor. Vast numbers of work animals, primarily cattle and buffalo are needed. They may also be utilized for food but the primary reason why they are kept is as a source of farm power. The reason for utilizing so much human labor has already been discussed. Pigs and poultry are kept as scavengers and are used as food by the human population. Fish are also taken from the irrigation reservoirs and canals.

Subsistence Crop and Livestock Farming is the other agricultural system found in a subsistence agriculture. It is found primarily in the poorer regions of Europe and involves small farms, each with a few head of livestock. It is also found in the colder, drier parts of Asia, the river valleys of the Middle East, Africa and Latin America. Most of the produce of the land and animals is utilized by the farmer with only a small amounts available for sale or barter. Self-sufficiency rather than profit is the objective of farmers engaged in this system of agriculture. The dominant crops are wheat, barley, millet, oats, corn, sorghum, soybeans and potatoes. Dual-

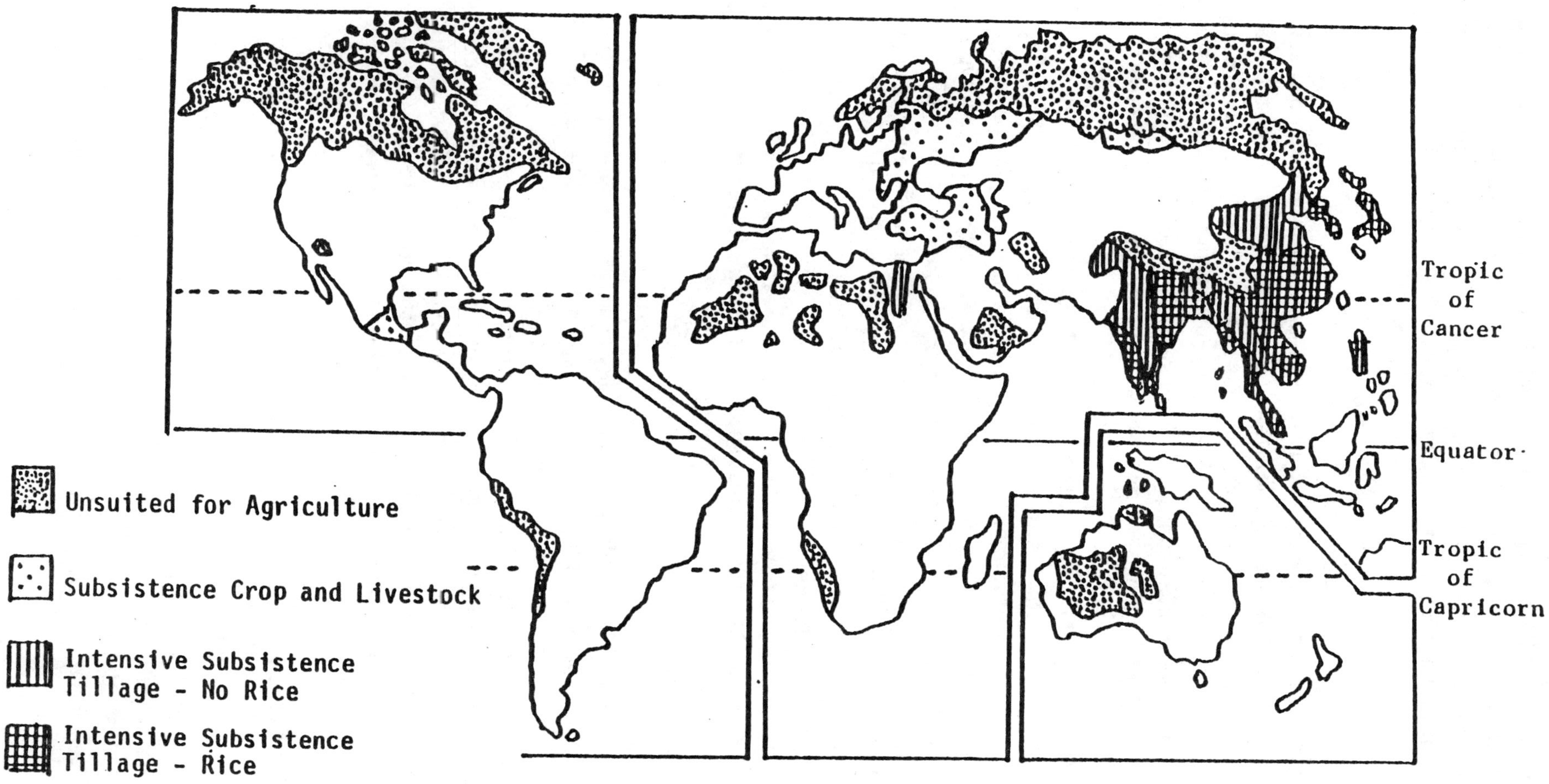

Figure 3. Worldwide Distribution of Agricultural Systems Found in Areas with a Subsistence Agriculture. (Adapted from Whittlesey, 1936).

purpose animals (work and food) are used. Milk is more important than meat. Cattle, pigs and sheep are the large animals usually raised. Chickens and ducks are often kept. In South America llamas and alpacas are raised. A cash crop such as cotton, flax, hemp or tobacco is raised to generate a meager income to purchase items they cannot grow.

Developed Agriculture

Figure 4 presents the distribution of the agricultural systems found in a developed agriculture. The major livestock producing systems are Ranching and Commercial Livestock and Crop Farming. However, for the sake of completeness, a discussion of the other systems is included. Profit is the single most important objective of agricultural enterprises in areas with a developed agriculture. The desire to realize maximum profit influences the decisions as to what class or classes of livestock are produced and the specific system of production that is practiced.

Ranching is the commercial use of the dry areas of the world. In very general terms, ranching is the predominant system in areas with less than 20 inches of rainfall (though practiced in areas that receive as much as 25 inches of rainfall). Livestock is the primary enterprise in ranching. Generally, ranching is based on cattle or sheep, and only occasionally are goats involved. The management is very extensive. Usually, only one species is involved in a given ranching enterprise. However, mixing species will often give better use of the land and is being practiced more and more. Unlike the nomads, ranchers are permanently located on either land they own or lease and very definitely seek to make a profit from their livestock. While primarily concerned with livestock production, some farming may be practiced, if possible, to produce forage for winter feeding. Most cattle ranches are found in the US, Canada, tropical and subtropical Latin America and the hotter parts of Australia. Most sheep ranches are in Australia, New Zealand, South Africa and Argentina. Regardless of where they are found, ranchers are usually of European ancestry and represents a holdover of the European colonial system.

Commercial Crop and Livestock Farming predominates in areas with more than 30 inches of rainfall but is also practiced in areas that have 25-30". Crop production is the primary enterprise with livestock secondary to crops. Beef cattle, sheep, dairy cattle, hogs, horses or poultry will be found as secondary enterprises on these farms. Often, animals are finished for slaughter in the winter when the farmer is not working the crops. They may specialize in only one species or be diversified, raising two or more. The production is usually intensive. This type of farm is on the decline in much of the developed world. This is what most would have historically referred to as the Family Farm. As mechanization and large unit ownership increases in agriculture, these farms are disappearing.

Commercial Livestock Finishing is a modification of Commercial Crop and Livestock Farming that has only been in existence for the last half of this century. The Corn Belt of the US, the Plains area of the US, parts of western and central Europe and now some in Australia, Brazil and South Africa are the major areas. The US is the dominating practitioner of this agricultural system. This represents a further specialization of agriculture. Generally, these producers don't raise any crops, rather they buy crops to feed their livestock. This modification is why they no longer fit into the Commercial Crop and Livestock Farming category. Their product is meat

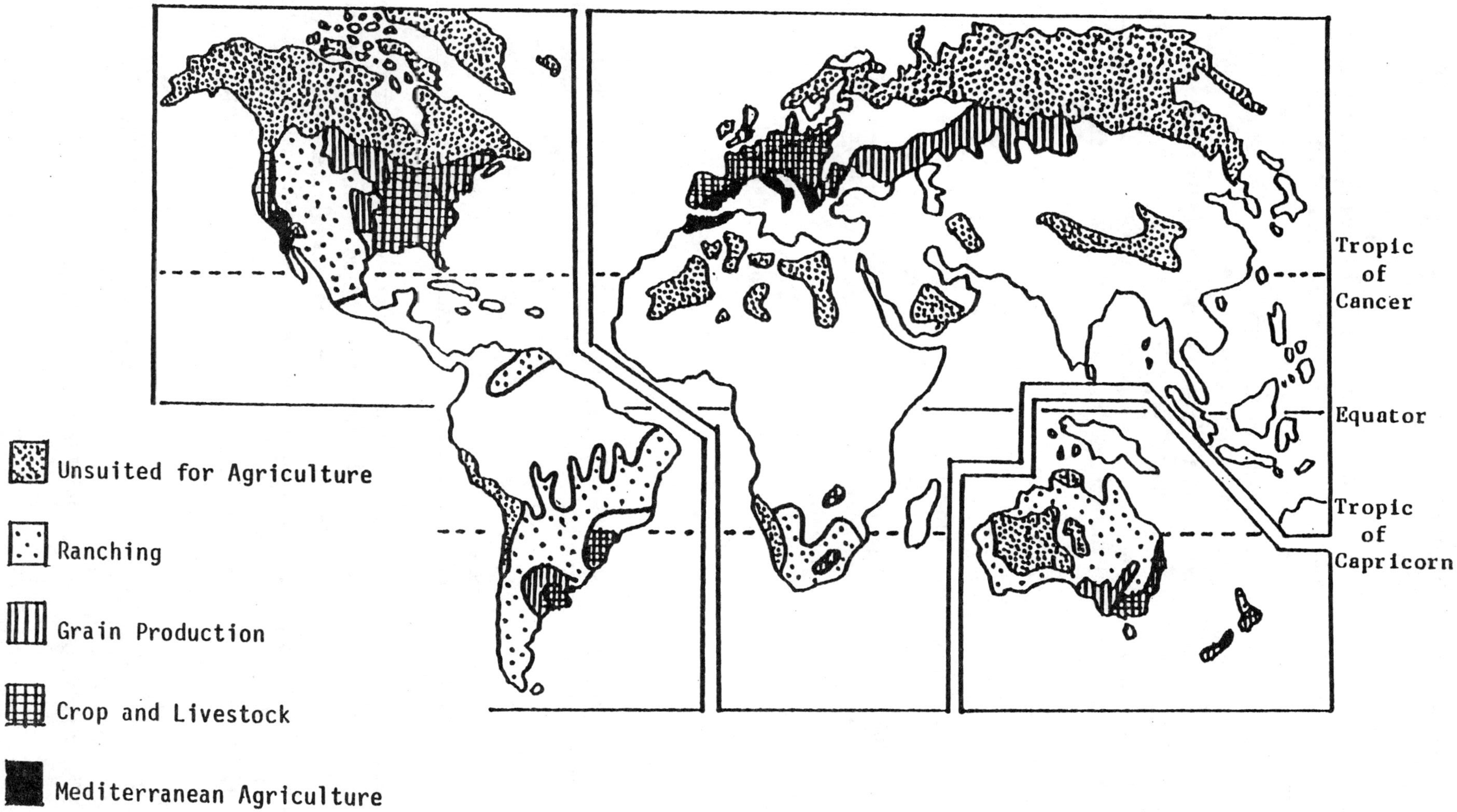

Figure 4. Worldwide Distribution of Agricultural Systems Found in Areas with a Developed Agriculture. (Adapted from Whittlesey, 1936).

animals ready for slaughter. Beef is the predominant type of animal fed in this system but some lambs are found also.

Commercial Dairy is also a modification of Commercial Crop and Livestock Farming. In many ways, modern commercial dairies resemble commercial livestock finishing operations. In the large dairy belts of the northeastern United States, western and northern Europe, southeastern Australia, and northern New Zealand, dairying depends on maximizing the use of pastures. In the colder areas, some land must produce winter feed especially hay and silage. The primary dairy products vary from region to region, and are largely dependent on how close the farmers are to their markets. Dairies near large population centers usually produce fluid milk, which is perishable and costly to transport. They also tend to operate year around. Dairies far removed from dense populations specialize in butter, cheese, or processed milk. As an example, New Zealanders, far removed from world markets, produce mainly butter which can be kept fresh and shipped to population centers. They are also much more likely to have a common, uniform dry period for the cows and not milk year around. In recent decades, a rapidly increasing number of dairy farmers have adopted confinement drylot systems rather than pasture systems and now raise their cattle on feed purchased from other sources or which they grow, harvest and bring to the cows (green chop). This is similar to livestock finishing. These are especially common in the western United States and near large population centers. Land near these centers is too expensive to graze and easy access to the population provides a good market. Even replacement females for their herd may be purchased. If they don't purchase replacements, they frequently contract with someone to grow them. In these large-scale operations, the number of cows is much greater than on family-operated dairy farms (250 vs. thousands) and hired labor is essential.

Commercial Grain Production is also in this category. It is practiced in areas receiving between 20 to 30 inches of rainfall. Some livestock may be produced but not much. The animals are usually stocker cattle and/or lambs brought in to graze small grains pastures in the fall and winter or beef cow/calf pairs used to clean up crop residues. The major emphasis is on producing small grains, usually wheat, or on irrigating summer grains. Australia has very large wheat belts as do the North American plains, the steppes of Russia and the Pampas of Argentina where this is found. Together, these places produce 45% of the world's wheat. Rice farms can be found in the USA in Arkansas, California, Texas and Louisiana.

Mediterranean Agriculture is an agriculture which is very different today than it used to be. In ancient times this was a peasant subsistence agriculture. It was based on small grains (wheat and barley) for rainy winter season crops, drought resistant trees and vines (grape, olive and fig) for the summer along with sheep, goats and some pigs. The grain was often raised in the orchards. The animals weren't really integrated into the land. They did not raise feed, collect manure or keep draft animals. The animals were kept in the mountains, the crops in the hills and valleys below. However, irrigation has changed all this. Today much of this region is referred to as Market Gardening (see below) and is a commercial area. Some animals are still grazed in the mountains but citrus is now a dominant crop. Mediterranean agriculture is found in a relatively restricted part of the world. (Where would that be?) Grapes are an important crop and these areas are famous for their wines. These areas are usually mountainous, and often use mountain pastures by what is called transhumantic practices. This practice involves slowly grazing livestock up the mountains during the summer. They reach the highest elevations by mid-summer, then graze slowly back down the mountain to reach the valley before snow falls. This differs from Nomadic Herding because transhumantic herders do not move their families and belongings with them. Transhumantic practices are not restricted to Mediterranean Agriculture but are practiced in the Alps and other mountains areas of Europe and the Western USA. In the U.S. cattle are usually released in the spring and rounded up in the fall with little herding in between.

Market Gardening (Truck Farming). This developed agriculture has developed in response to large urban centers. They have no livestock and specialize in fruits, vegetables and vine crops. The entire output of the farm is raised for sale. They frequently depend on migratory seasonal labor. Scattered bands of this can be found throughout developed countries.

Plantation Agriculture. This agriculture is another holdover of the European colonial system. Large tracts of land are devoted to the culture of tropical or sub-tropical crops such as the fiber crops, (Hemp, sisal, jute or cotton) or luxury, non-essential crops like tobacco, cacao, coconuts, tea, various spices, coffee, sugarcane and bananas. These are found in developing countries and the products are grown for export to developed countries for profit. Much of the labor is hand labor and some draft animals are used. Odds are your coffee for your morning cup or tea for the cooling glass of iced tea were once hauled to market or from the field on the back of a donkey.

Developed vs. Subsistence Agriculture

In comparison with the agriculture of the developed countries, subsistence agriculture appears to be very inefficient and unproductive. However, it is a serious error to judge all economic systems in terms of what is considered to be the best in a developed country such as the United States or Western Europe. The reason for this statement is explained very well in a paper presented in 1975 by Dr. James E. Johnson which evaluated the traditional agricultural systems found in Asia. The following are the first two paragraphs of this paper.

> "The traditional agricultural systems of Asia have developed over a period of thousands of years. They are basically subsistence systems requiring few external inputs but designed to provide for the essential food, fiber and shelter requirements of the farm family plus a small surplus to be used as emergency reserve, to pay taxes and for sale or barter. They normally include a balance of crop and animal production in which the animals contribute labor (cattle and/or water buffalo), manure and food (milk, eggs and meat) but seldom consume any food which would normally be consumed by the human population. These balanced systems are still prevalent throughout Asia but are being eroded by diminishing size of land holdings, rapid urbanization with concomitant demands for greater food surpluses as well as the introduction of western concepts of monoculture and intensive animal production systems. Nevertheless most of the animal products now consumed in the developing countries of Asia are still produced by traditional type systems."

> "The traditional Asian systems are frequently criticized as inefficient. The animals are usually slow growing, late maturing and smaller in size and production of milk, eggs and meat than western counterparts. We must look carefully at our definition of efficiency before reaching this conclusion, however. The western concept that the most efficient animal is the one that grows fastest or that produces the most eggs, milk or meat per unit time does not necessarily hold in the Asian context. These systems are based on low capital investment in animals, little or none in housing, use of labor with no alternate options and of feedstuffs which have little or no other utility. This is not to say that they could not be improved by additional marginal inputs, but they have developed in such a fashion as to maximize the efficiency of use of available land, labor and nutrients under existing systems...."

There are several points in the second paragraph of Dr. Johnson's paper that deserves further attention. It is difficult to overemphasize the importance in subsistence agriculture of the low capital investment in animals and housing for the animals and farm equipment. He could also have included farm implements and other equipment. The majority of the farmers have a low income, so there is little, or no, money available to finance capital improvements of any kind in the farm enterprise.

The farm implements and machinery used in the developing countries appear to be very crude and inefficient in comparison to that used in a developed agriculture. This is not really a fair comparison. They have several advantages. They have the advantage of a low initial cost and a low maintenance cost. Of even greater importance, when repairs are needed they can usually be done locally using relatively cheap, readily available materials and with a minimum of delay.

There are several reasons why animals rather than tractors are the predominant source of power used in subsistence agriculture. Obviously, they cost less, both initially and to replace. They are a renewable resource that can be produced locally, usually by the farmer himself. Machines require fossil fuels which are both expensive and usually in short supply. The fuel for animals can be produced by the farmer, largely consisting of plants that cannot be consumed by humans.

The final point has to do with the use of labor that has no alternate option. This is a major contribution of agriculture in Asia with its vast human population. It provides work for a vast number of people that would otherwise be unemployed. This labor pool will continue to be utilized in agriculture by practicing labor intensive systems as long as human labor is more economical than machines. If the time ever comes when there are non-agricultural demands for the service of these people they will no longer be a cheaper source of labor than are machines and they will be replaced in agriculture by machines. Until such time, however, the utilization of vast amounts of hand labor is not inefficient.

Comparison of the Livestock Industry in the Developed and the Developing Countries

Table 3 presents a summary of a comparison of the livestock industry in the developed and developing countries. The two are comparable in that each has a very few units that are very large in size, but differ in that such units in the developed countries are usually much more productive. However, even though not managed as efficiently as in the developed countries, the very large units are usually the only livestock producing units in the developing countries that utilize any modern technology or have improved breeds of livestock.

The two groups of countries differ radically in numbers of livestock units that are moderate to large in size. Such units make up the majority of the units in the developed countries, but are almost totally absent in the developing countries. In the developed countries, these are the farms and ranches owned and operated by the middle class and this is the class that is almost nonexistent in the poor countries of the world.

By far the most numerous type of livestock enterprise in the developing countries are small units. Small units are also fairly numerous in the developed countries but for a different reason. In the developed countries, the small units are seldom the major source of income for the owner/operator. In sharp contrast, almost all of the small units in the developing countries are not only a major source of income, they are usually the only source of income for their owners.

As shown in **Table 3**, the small units in both groups of countries are similar in: (1) they utilize a limited land area; (2) they are operated almost entirely by the labor of the family; and (3) cash returns are low. The most important reason for low cash returns is the small size of the operation, however, returns per animal unit are usually less than is true in larger units. In the developed countries one of the main reasons for this is the small producers do a poor job of marketing their animals even though good markets are available. In the developing countries low returns are the result of the combined effect of the inavailability of good markets and a general low level of animal productivity. Usually the only cash market available to the producers in the developing country is from traders that offer a very low price and the producer has no choice because they are the only buyers. The reasons for the low level of animal productivity are given in **Table 3.**

The small units in both groups of countries are likely to remain small, but for different reasons. In the developed countries, small units are usually a sideline and many are a hobby rather than a business. The operator has a job in town but lives on a few acres in the country. He keeps a few animals because he has the space to do so and enjoys working with animals but he has no desire to expand. In contrast, most small farmers in the developing countries would like to expand but have little, or no, opportunity to do so for the reasons given in **Table 3.**

Two additional points should be made about the livestock industry in the developing countries: (1) livestock are seldom kept primarily to produce food; and (2) animal performance is usually submaximal from a biological standpoint so there is room for improvement and still stay within the system that has developed.

3903chap5Adoc

Table 3. Comparison of the Livestock Industry in the Developed and in the Developing Countries.

Size of Unit	Characteristics of the Livestock Industry Developed Countries	Developing Countries
Very Large	Very few	Very few
Moderate to Large	The majority	Almost none
Small	Fairly numerous	By far the most numerous type of livestock enterprise
	Usually not the major source of income for the owner/operator	The major source of income for the owner/operator
	1. Utilize limited land area 2. Operated almost entirely by family labor 3. Cash returns to livestock enterprise are low as a result of:	
	a. Small size b. Inefficient marketing	a. Small size b. Poor markets available c. Low level of animal productivity 1. Genetically poor animals 2. Low nutritional levels 3. Diseases and parasites
	Little or no desire to expand 1. Livestock usually more of a hobby than a business	Little opportunity to expand 1. Total income needed for subsistence, none available for expansion. 2. Little, or no, collateral for loans if loans were available

References

Baker, P.R. 1978. Nomadism in Africa in G. Williamson and W.J.A. Payne. An Introduction To Animal Husbandry in the Tropics, 3rd Ed., Longman, London and New York.

FAO. 1994. Production Yearbook. FAO, Rome.

FAO. 1981. The State of Food and Agriculture. FAO, Rome.

Harris, M. 1974. Cows, Pigs, Wars and Witches. Random House. New York.

Jackson, M.G. 1981. A new livestock development strategy for India. World Ani. Rev. 37:2.

Johnson, Douglas L. 1969. The Nature of Nomadism: A Comparative Study of Pastoral Migrations in Southwestern Asia and Northern Africa. Univ. of Chicago Dept. of Geography Res. Paper No. 118.

Johnson, James E. 1975. Conflict and Complementary in Asian Animal Feeding and Food Production. Prepared for Rockefeller Foundation Review of the Role of Animals in the Future World Food Supply, New York, April 21-22, 1975.

Jordan, T. G., M. Domesh, and L. Rowntree. 1994. The Human Mosaic: Thematic Introduction to Cultural Geography, 6th ed. Ch. 3, The Agricultural World, p.86, Harper Collins College Publishers.

McDowell, R.E. 1978. Are we prepared to help small farmers in developing countries. J. Animal Sci. 47:1184.

Waters, R.F. 1971. Shifting Cultivation in Latin America. FAO Forestry Development Paper No. 17. FAO, Rom.

Whittelsey, D. 1936. Major agricultural regions of the earth. Annals. Assoc. Amer. Geographers 26:199.

Study Questions

1. Briefly (I mean brief - ¼ page, handwritten) describe each of the systems of agriculture discussed in this chapter and tell in which level of agricultural development each is found.
2. Explain how climate (particularly rainfall) affects which agricultural system is practiced in a given level of agricultural development.
3. Which 2 systems of agriculture are the only two that are concerned almost exclusively with livestock?
4. What are transhumantic practices?
5. What is the most common reason why a given area of land will be devoted to grazing of livestock? What are 3 other conditions that, when present in an area, usually means the land will be devoted to grazing?
6. In what geographical area is most of the intensive subsistence tillage found, and why is rice grown whenever possible?
7. Describe the traditional agricultural systems of Asia. List 3 different advantages to the developing countries of the relatively crude farm implements and machinery that they use even though they are inefficient compared to those used in highly developed countries. List 3 reasons why animals rather than tractors are the predominant source of power used in subsistence agriculture. Explain why so much human labor is used in Asian agriculture.
8. What are the geographical regions of the world where primitive agriculture predominates?
9. What is the major difference between the developed and developing countries as far as numbers of the different sized livestock producing units are concerned? Why is this so? (This refers to very large, moderate and small sized units.)

Chapter 6

The Agricultural Animals That Serve Human Societies

Introduction

Man's progress on the planet has been due in large measure to the animals he has used for millennia. We have reached the point in this text where we will start considering some of them and their uses more specifically. We begin that process by taking a broad look at which animals comprise the domestic species and which are referred to as agricultural animals. It is a short list, since only a relative handful have been successfully domesticated. One has to wonder why no more have been domesticated considering that there are millions of animal species in the world. No large animal species has been domesticated for at least 4000 years. The reasons for this must surely be found in the individual natures of the species themselves though that seems hardly enough to explain it fully. Perhaps there will be more successful domestications in the future.

Domestication and Domination

All agricultural animals are domesticated animals, which means: (1) they have been brought under human control; and (2) they have adapted to symbiotic life with their human owners. It is not uncommon to find the term domesticated animals shortened to domestic animals.

It is important to distinguish between the animals that are truly domesticated and those that are basically wild but which have been brought under human domination. Among the latter group would be zoo animals at one extreme and animals that have been tamed at the other extreme. The prime example of the latter group are elephants, which are tamed and widely used as work animals in India. While some of the work they do is similar to that done by agricultural animals, elephants are tamed rather than domesticated animals and will not be included as an agricultural animal in this book.

In order to adapt to symbiotic life with the human there are three different things animals must be able to do: (1) they must be able to adapt to a fairly wide range of physical environments; (2) they must be able to adapt to the "artificial " environment that the human imposes on them; and (3) they must possess the capacity to respond to selection for some specialized functional development considered important by their human owners.

It should be apparent why, if a given species is to be very useful under domestication, it must be able to adapt to wide range of environments. The human lives in a wide range of physical environments and, to be of service to him, the animals must also be able to survive in these environments. Any stress that is not a part of the physical environment, but rather is imposed by the human, is a stress of the "artificial" environment. Included are things such as: placing normally free ranging animals in confinement; confining ruminants in a pasture that contains only one species of plants when in nature the animals would seek out and graze a variety of plants; altering the metabolism of animals by implants, feed additives, etc.

Animal "Improvement"

Livestock producers spend much effort in an attempt to "improve" domestic animals to produce more of a product or service of benefit to the producer by selection to modify or improve specialized functional development of the animals in: (1) locomotion; (2) rate and type of growth and fattening; (3) mammary development and increased milk production; (4) body covering; or (5) in some other characteristic considered to be of value such as for use in sport (hunting, bull fighting, etc.), as a pet, as a laboratory animal or as a status symbol. Generally, the more the animal is changed by selection to "improve" them for a particular specialized functional development, the less adapted the animals become for survival in the wild, and the more dependent they become upon their owners.

The specific type of specialized functional development involved with locomotion depends on whether the animal must pull or carry the load and whether speed or strength is the most important. The animal that comes first to mind is the horse or its hybrid, the mule. However, in various parts of the world the beast of burden may be cattle, water buffalo, yak, camel, llama and even sheep, goats or dogs. Indeed, in some areas of the world the most important use made of cattle is for work.

One of the primary uses of domesticated animals in human societies is as a source of food. If they are primarily utilized for meat, then the emphasis is on how rapidly they produce the meat, the quality of the meat and the amount, quality and distribution of the fat. No one would question the statement that since domestication, many profound changes have been made in the meat animals in "improving" them as a meat producer. If the primary food product yielded by the animals is milk, the emphasis has been on increasing the yield. This has been accomplished, but in the process has created an animal much less capable of survival in the wild. Imagine the difficulty in an African herd that a modern high producing Holstein, with her huge udder, would have in running fast enough to escape from a lion. Finally, if the primary product is eggs, increased production has required a complete change in the normal reproductive physiology of the hen. She has been selected to produce eggs out-of-season, that is during the seasons when wild birds are completely inactive sexually and the modern hen will no longer brood eggs to hatch young chicks.

The Domesticated Animals

There are 33 domesticated species of animals that are generally considered to be of economic importance.

- Two are insects (bee, silkworm).

- Five are birds whose economic importance is based on their aesthetic value rather than because they produce a useful product (swan, guinea fowl, pea fowl, pigeon, ostrich). Four birds are important as food producers (chicken, turkey, goose and duck).

- Seven are small mammals (dog, cat, rabbit and the laboratory species - rats, mice, hamsters and guinea pigs).

- To agriculture 19 species are of the greatest importance. These include four species of birds listed above (hen, turkey, duck and goose), plus 15 mammalian species of which 12 are ruminants.

Of the ruminant species, the three that are most widespread in distribution are cattle, sheep and goats, with cattle and sheep by far the most numerous. At least a few of each are found in every country of the world. Two species that are very limited in distribution, but very important in the areas in which they are found are the water buffalo and the camel. The following ruminant species are extremely limited in their distribution, but are also very important in the areas where they are found: llama, alpaca, yak, banteng, reindeer, musk ox, and the gayal. There are three non-ruminant (monogastric) mammalian species: the horse, ass and pig. Unlike the pig, which cannot utilize roughages, the horse and ass have a functional caecum which permits them to utilize roughages such as grass and hay. The caecum in the pig (as in the human) is a non-functional rudimentary organ located at the junction of the small intestine and large intestine. In the human it is called the appendix, and about the only time we are concerned with it is when it gets inflamed (appendicitis). In the horse and ass (and a few other species, such as the rabbit and guinea pig), the caecum is functional. It contains bacteria which function in the same way as the bacteria in the rumen in breaking down complex carbohydrates into volatile fatty acids that can be absorbed and utilized by the animal as a source of energy.

Several other species of animals are also used by various peoples of the world and some are agricultural animals. Others are wild species regularly harvested by people for various reasons. The major ones of these are in **Table 1.**

Animal Use - General Considerations

If the average person was asked to list the major uses human societies make of their agricultural animals, there are three major categories that would probably be listed: (1) for food; (2) for their body coverings; and (3) as a source of power. This list would be complete if one restricted it to the countries with a developed or high subsistence agriculture. However, when countries with low subsistence or primitive agriculture are considered, there not only are other uses, but the other uses may be the most important reason for keeping agricultural animals in a given society.

In very general terms, only the first two categories listed in the previous paragraph apply in areas with a well developed agriculture. Furthermore, in these areas the animals are usually kept on a given unit to primarily produce one product only. For example, sheep produce both meat and wool. However, in a country with a developed agriculture the sheep are usually raised primarily for meat with wool secondary if there is enough forage to insure weaning a high percentage of fat lambs ready for market. In such flocks, the mutton breeds are used. If forage is not ample, wool is emphasized with meat production secondary and the fine wool breeds are used. Cattle are

Table 1. Non-traditional animals and some fowl used for meat in certain areas of the world.

Name	Habitat
Agouti	Gulf of West Africa
Alpaca	S. America Andes mountains
Angora goat	U.S., South Africa, Nambia
Anoas	Southeast Asia (related to buffalo)
Babirusa	Southeast Asia (related to pig)
Bateng	Southeast Asia
Bateng-cattle hybrids	Southeast Asia, Indonesia, Malaysia
Bearded pig	Southeast Asia (related to pig)
Bison	U.S., Europe
Buffalo	Egypt, Italy, USSR, China Southeast Asia
Camel	Africa, Asia
Cape Buffalo	Central, Eastern and Southern Africa
Carabao	Southeast Asia (swamp buffalo)
Capybara	S. America, Upper Amazon Basin
Caribou	Arctic Region, U.S., Europe, Asia
Deer	
White-tailed	Canada, U.S.
Black-tailed	Canada, U.S.
Mule deer	Canada, U.S.
Axis	U.S.
Roe	Europe, USSR
Red	Europe, USSR
Reindeer	Canada, Scandinavia, USSR
Duck	All over (> 700 types)
Ėland	Central, Eastern and Southern Africa
Elephant	Central, Eastern and Southern Africa, Asia
Elk	Canada, U.S.
Gaur	Southeast Asia
Giraffe	Central, Eastern and Southern Africa
Grant's gazelle	Central, Eastern and Southern Africa
Guinea fowl	S. America, Africa
Guinea pig	Ecuador, Peru, Chile, Bolivia
Horse	All over
Impala (antelope)	Central, Eastern and Southern Africa
Kangaroo	Australia
Kongoni	Central, Eastern and Southern Africa
Kouprey	Indo-China (ruminant)
Llama	S. America Andes mountains
Moose	Canada, U.S.
Oryx	Central, Eastern and Southern Africa
Pigeon	Europe, Africa
Pronghorn antelope	U.S.
Rabbit	All over
Tamaran	Southeast Asia (related to buffalo)
Thomson's gazelle	Central, Eastern and Southern Africa
Wildebeest	Central, Eastern and Southern Africa
Yak	Himalayan mountains
Yak-cattle hybrids	Himalayan mountains
Mithan	Southeast Asia
Sulawesiwarty pig	Southeast Asia
Pygmy hog	Southeast Asia

Source. McDowell. p. 28-29. Used with permission.

either beef cattle kept to produce only beef, or dairy cattle kept primarily to produce milk with meat production definitely secondary.

In areas of subsistence or primitive agriculture, the agricultural animals are kept to produce as many products as possible for their owners. Thus, strains or breeds of animals found in these areas produce two or all three of the above products, and are referred to, respectively, as dual purpose or triple purpose. An example would be dual purpose cattle producing both milk and meat, and triple purpose cattle that also are used for work animals.

Thus, not only the way in which animals are utilized, but the degree to which they are utilized, is very dependent on the level of agriculture of a region. The higher the level of development, the more the emphasis is on profit and the more specialized animal production becomes. A herd or flock is primarily kept to produce a product to sell rather than one to be utilized by the owner. The lower the level of agricultural development, the more likely it is that the animals are kept to produce a product, or products, for the use of the owner, and are likely the only way the owner would be able to obtain these products. Hence, the more important it is that the animals be dual purpose or triple purpose in these areas.

If two species are equally well adapted, the most important factors that determine the choice are usually socioeconomic. Economic factors are the most important in areas with a highly developed agriculture. In other words, the choice is usually the species that will yield the greatest profit. On the other hand, social factors, primarily those related to culture and tradition, are more important in areas of less highly developed agriculture. In primitive agriculture, factors related to culture and tradition may be the major determinants. For example, in many areas of Africa, raising either goats or sheep would contribute more to the nutritional well-being of the people of an area. Yet, instead of goats or sheep, cattle are kept in large numbers strictly for the prestige that large herds of cattle accrue to their owners.

In subsequent chapters we will examine in more detail the specific uses of these animals as well as some of the specifics of how they are used in different societies. You may find your notions of animal use stretched as we examine those uses. Most people in developed countries have a bias that tends to prevent their objective evaluation of world livestock production. Citizens of the First World countries are taught to worship productivity. The best animal is the one that grows the fastest, or which produces the most of some product, particularly if it does so efficiently. For this reason, we often adopt a rather patronizing attitude towards animal agriculture in the developing nations. By American standards, their animals are lowly productive and very inefficient. We lose sight of the fact that often these animals utilize what would otherwise be waste and eventually convert it to human food. They may do this at a slow rate but they are never in competition with man for the available food.

Understanding this idea is probably the single most important aspect of this entire text. These ideas have been touched on in earlier chapters and will be again before this book is ended. Our way (whoever we are) is not always the best way for others (whoever they are). Understanding this is absolutely essential if you are to truly understand world animal agriculture and the people involved with it.

References

Bhattacharya, P. 1978. Buffalo. Chapter 6 in Williamson, G. and W.J.A. Payne. An Introduction to Animal Husbandry in the Tropics, 3rd. Ed. Longman. London and New York.

Devendra, C. 1978. Goats. Chapter 8 in Williamson, G. and W.J.A. Payne. An Introduction to Animal Husbandry in the Tropics, 3rd Ed. Longman. London and New York.

FAO. 1994 FAO Production Yearbook.

Fernandes-Baca, S. 1978. Llamoids or New World Camelidae. Chapter 10 in Williamson, G. and W.J.A. Payne. An Introduction to Animal Husbandry in the Tropics, 3rd Ed. Longman. London and New York.

Grizimek, B. 1968. Grizimek's Animal Life Encyclopedia. Ch. 13. The Wild and Domestic Oxen by A. Wunschmann.

Johnson, James E. 1975. Conflict and Complementary in Asian Animal Feeding and Food Production. Prepared for Rockefeller Foundation Review of the Role of Animals in the Future World Food Supply, New York, April 21-22, 1975.

McDowell, R.E. 1991. A Partnership for Humans and Animals. Kinnic Publishers. Kinnickinnic Agrisultants, Inc., Raleigh, N.C.

USDA. 1975. Foreign Agricultural Service. World Agricultural Production and Trade. Statistical Report. May 1975.

STUDY QUESTIONS

1. Describe the difference in a domestic animal and a dominated animal. Give an example of each.
2. What are some things that man does which can cause stress to a domestic animal?
3. Why is it that the more an animal is "improved" for agricultural purposes the less likely it is to survive in the wild.?
4. Discuss the differences in a dual purpose animal and a triple purpose animal. Give examples.

Chapter 7

The Food Uses of Agricultural Animals by Human Societies

Introduction

Food, one of man's basic needs, has been provided by animals since long before man began farming and turned to domestic animals to supply their needs. Evidence indicates that man progressively moved from scavenger to hunter of wild game and then to settled agricultures and domestic species. With domestication came great advancement in man's culture. No doubt, the readily available food supply provided by domestic species set the stage for civilization to emerge. In this chapter, we will examine the food contributions of agriculture animals.

Meat Production

The nutrients provided by various meats, including that from fowl and marine life, are important globally. Protein and energy are especially important. Annual per capita intake ranges from over 230 lb. per year in affluent countries to little or none in semi-vegetarian societies. In developing countries, annual per capita consumption is estimated at 20-45 pounds. This is probably an underestimation because contributions from marine life and exotic species found in specific localities are not included. **(Table 1).**

World meat production has been more or less increasing steadily for many years at a rate of something over 3% per year. Total world production increased by 44% from 1979 to 1994. Meat production from the major meat-producing species is given in **Table 2.** The pig is the most important meat producer, producing 40% of the world's meat. Beef is next 27%, followed by poultry with approximately 25%. Together, these three sources produce 92% of all meat from domestic species.

Poultry meat production is growing the most rapidly at 5-6% per year. Poultry meat production has nearly doubled in the last 15 years. Some countries such as China are undergoing absolutely phenomenal production increases. They are currently producing 3 fold as much poultry meat per year as they were 15 years ago. Developed countries are also increasing production. This increase is in response to two things primarily. First, the technology, genetics and management know-how associated with the poultry industry has made its production much more cost effective and therefore the product cheaper. Second, economic development in many countries has increased incomes and made a better diet more available to more people. Time and time again people demonstrate their desire to purchase more meat for their diet as soon as they are able. Goat meat production is also increasing 5% per year. Pork production is increasing at the rate of 3-4% per year. Even buffalo meat is registering yearly increases in production. Mutton, lamb and beef are growing more slowly, perhaps registering as much as 1% increase in some years. Horsemeat production is declining.

Table 1. Non-traditional animals and some fowl used for meat in certain areas of the world.

Name	Habitat
Agouti	Gulf of West Africa
Alpaca	S. America Andes mountains
Angora goat	U.S., South Africa, Nambia
Anoas	Southeast Asia (related to buffalo)
Babirusa	Southeast Asia (related to pig)
Bateng	Southeast Asia
Bateng-cattle hybrids	Southeast Asia, Indonesia, Malaysia
Bearded pig	Southeast Asia (related to pig)
Bison	U.S., Europe
Buffalo	Egypt, Italy, USSR, China Southeast Asia
Camel	Africa, Asia
Cape Buffalo	Central, Eastern and Southern Africa
Carabao	Southeast Asia (swamp buffalo)
Capybara	S. America, Upper Amazon Basin
Caribou	Arctic Region, U.S., Europe, Asia
Deer	
White-tailed	Canada, U.S.
Black-tailed	Canada, U.S.
Mule deer	Canada, U.S.
Axis	U.S.
Roe	Europe, USSR
Red	Europe, USSR
Reindeer	Canada, Scandinavia, USSR
Duck	All over (> 700 types)
Eland	Central, Eastern and Southern Africa
Elephant	Central, Eastern and Southern Africa, Asia
Elk	Canada, U.S.
Gaur	Southeast Asia
Giraffe	Central, Eastern and Southern Africa
Grant's gazelle	Central, Eastern and Southern Africa
Guinea fowl	S. America, Africa
Guinea pig	Ecuador, Peru, Chile, Bolivia
Horse	All over
Impala (antelope)	Central, Eastern and Southern Africa
Kangaroo	Australia
Kongoni	Central, Eastern and Southern Africa
Kouprey	Indo-China (ruminant)
Llama	S. America Andes mountains
Moose	Canada, U.S.
Oryx	Central, Eastern and Southern Africa
Pigeon	Europe, Africa
Pronghorn antelope	U.S.
Rabbit	All over
Tamaran	Southeast Asia (related to buffalo)
Thomson's gazelle	Central, Eastern and Southern Africa
Wildebeest	Central, Eastern and Southern Africa
Yak	Himalayan mountains
Yak-cattle hybrids	Himalayan mountains
Mithan	Southeast Asia
Sulawesiwarty pig	Southeast Asia
Pygmy hog	Southeast Asia

Source: McDowell, 1991. p.78-79.

Table 2. Meat Produced by the World's Major Species

	Slaughtered (1,000 Head)			Meat Produced (1,000 MT)			
	1979-81	1992-94	1992-94 as of percent of 1979-81	1979-81	1992-94	Rank	1992-94 as a percent of 1979-81
Beef	229,983	239,508	104	44,938	50,622	2	113
Buffalo	11,811	18,024	153	1,542	2,469	6	160
Sheep & Lamb	382,026	462,951	121	5,632	6,911	4	122
Goat	157,079	244,787	156	1,714	2,923	5	171
Pig	742,132	986,983	133	51,935	75,642	1	146
Horse	-	-	-	525	513	7	98
Poultry	-	-	-	25,949	46,912	3	181
Misc.	-	-	-	2,846	3,147	-	110
Meat Total				135,081	189,139		140

Most meat comes from the species discussed above. In addition approximately 1.0 billion animals of about 50 species **(Table 1)** each provides up to 5,000 MT of meat yearly. Most of these animals are geographically limited but are very important to some peoples.

The Value of Animal Products in the Human Diet

Most of the human population is omnivorous. Even many who call themselves vegetarians still consume dairy products and eggs. In wealthy societies, meat is the food item that the meal is built around. The meat is chosen first, the rest of the meal is selected to complement the meat. According to McDowell, 1991, p29, "Among developed country societies, 32-38% of total food expenditures are for meat and fish. This level appears consistent irrespective of income level. The highest income groups in developing countries also use over 30% of their total food expenditures for meat, fish and dairy products..."

Due to economic status and lack of availability, meat is not the main dish for the world's low income groups. However, these groups seek to add as much meat as possible to their diet to supplement the bland taste of the starchy foods which of necessity form the bulk of their diets. Low income groups are not able to spend as much as the higher income groups (only 5-15%). However, these groups identify meat, fish, fluid milk and fermented milk as essential food items even though they must restrict their purchases.

Whether the people who consume meat are educated or not, there is generally a very poor understanding of the physiological reasons why we eat meat. Meat is an important source of minerals (calcium, phosphorus and iron), protein and essential vitamins, especially vitamin B_{12} as well as calories. It is a very important part of diets for the human population because it is such a high quality food. This is illustrated below.

Sorghum grain is a staple food for many of the world's people, especially in Africa. Sorghum has enough total protein to supply adult humans with their needs if it were all available. Unfortunately, part of the protein is not available for people to digest because of the presence of the compound tannin found in sorghum. Tannin is a problem because it negatively affects the digestion of the protein in sorghum. The logical solution might be to reduce the amount of tannin in sorghum (by selective breeding), however, tannin helps make sorghum bird and disease resistant. Plant breeders are actually developing varieties with more tannin. The net effect is more sorghum for human consumption but with less available protein. Adding some high quality animal protein, even two or three times/week to the human diet allows people to continue to consume high intakes of sorghum.

Other staple starch (calorie) crops such as rice, corn, wheat, barley, root crops (cassava, yams, potatoes), and tree crops (plantain) are also too low in protein for human diets. The use of small quantities of animal products permits higher use of these foods than if the protein came from beans or other food legumes. This is important because these starch crops are high yielding and thus supply many calories per acre. Example: Several West African countries have high dependence on sorghum, millet, cassava and corn as foods. They also have large animal populations to supply quality in the diet and they import additional supplies.

Absolute quantities of animal product needed in the diet to balance the plant-derived foods are small. A 2 kg (4.4 lb.) chicken will supply the total protein and methionine needs of eight children per day. One-third cup of milk, 25 grams of meat or one egg per day provides enough protein to allow a person to consume 1.0 to 1.4kg (2.2-3.1lb.) of starchy foods and still meet their protein needs. (McDowell, 1991).

A logical question which follows the above discussion is "Why not use plant proteins?" When plant proteins are used, cereals consumption must be reduced by 30-40% and the protein foods increased by that much to maintain a near balanced diet. Since yields per unit of land of high protein food crops are low, (particularly when grown on acidic soils common in these areas), balancing the diet forces the people to reduce yields from their fields. (McDowell, 1991) Another problem with food legumes, such as peas and beans, is that they are much more susceptible to loss in storage than the cereal grains. The essential point is that animal products in the human diet help maintain high intakes from the crops that can be best produced in many areas.

Since animal products are so important in feeding the world's population, one would think that much research would be directed at animal production. However, much more research is done on food crops than is done on animals. Even so, the contributions from animals have increased steadily, about 2% per year since 1955, which is about the same rate that food from food crops has increased. As pointed out by McDowell, 1991, "At present, animal products provide over 60 million MT of edible protein and more than 1 billion Megacalories of energy per year. Milk and milk products are the largest sources of both protein and energy followed by meat and eggs. With its high biological value, total protein from animal products is nearly equivalent to that from corn and wheat combined and more than half that from all cereals. The megacalories of energy are nearly equal to that from wheat and exceeds that from paddy rice." Nevertheless, animals rank well behind plants as a source for human food. On the whole, animal products supply 15-16% of the calories and 34-35% of the protein consumed by the worlds' people.

Meat Production and Agricultural Development

How does meat production fit into agricultural development? The lower the level of agriculture the <u>less likely</u> it is that the primary reason for raising <u>large</u> farm animals is to produce meat. Instead, the pig, chickens and small ruminants such as sheep and goats are kept as meat animals and cattle are raised for other purposes (primarily). A major reason for this is related to the lack of proper meat storage facilities. Most of these places are in hot areas of the world. A single family can consume a chicken, small pig, goat or sheep carcass before it spoils too badly to eat but not a cow carcass. The goat is often referred to as the poor man's refrigerator because of this. This is the reason that meat dishes in these areas are often very heavily spiced. The spices mask off-flavors of meat. Curry and peppers are often used this way. Cattle are used for meat but only in a salvage situation. Old, crippled or sick animals that are near death may be slaughtered or they are eaten after they die naturally. A cow or bull may be killed for a village feast in which it can all be consumed in a short time. Another reason is that the cattle are more important for some other purpose.

Milk and Milk Products

World milk production for the major species is given in **Table 3.** In the developed countries almost all of the milk is produced by cows. In low subsistence and primitive agriculture, other species make significant contributions. The water buffalo is the major one of these (especially in Asia) but sheep and goats are important as well. Even on these continents the cow is still the most important. On a world basis, approximately 88% of the milk is from cattle, 9% from buffaloes with the remainder from the following species: Alpaca, Ass, Bateng, Camel, Caribou, Elk, Goat (2.0%), Llama, Mare, Moose, Musk-ox, Reindeer, Sheep (1.5%), and Yak. Notice the pig is absent. Swine milk is not used anywhere to my knowledge.

Humans use about 526 million metric tons of milk per year in fluid form and prepared products. The average person in a developed country consumes 0.5 to 0.8 liter of milk (or equivalent) per day. In most developing countries, only 0.2 to 0.3 liter is consumed, however, some pastoral societies with high dependency on grazing animals are the highest consumers at 1.0 or more liters per day. (1 liter = 1.06 quart).

Milk is a very perishable product but is so valued that people have developed many techniques for preservation. The most common is cheese. Over 7,000 cheeses have been identified around the world. These are both fermented and non-fermented varieties. Milk fat is very valuable in developing countries. Butter produced by hand churning is used by the family or sold if there is a surplus. The main products are salted butter, ghee or butter oil, which are very valuable in cooking. To many people milk fat represents the major source of fat in the diet. In addition to adding flavor to high starch foods such as rice and root crops, the fats also provide the essential fatty acids in the diet and provide needed calories.

On the whole, the total production of milk in the world has been only increasing slowly for the last decade. However, Europe has declined significantly and the other continents have all increased. Asia has increased the most. Consumption of milk is declining some in developed countries, but expanding at the rate of 2 to 6% per annum in developing countries. Dairying is being promoted in the developing countries through the efforts of several major international lending agencies. Promoting a dairy industry assists in development several ways. In addition to improving nutrition by making available a very nutritious product, it also provides employment. Since dairying is a year-round job as opposed to crops that have seasonal labor needs, the labor is needed year around. This serves as an added means of income and bolsters the overall economy. Dairying is also a very efficient means of converting animal feed to people food. One kilogram of grain (2.2 lb.) fed to a dairy cow, along with roughage, crop wastes and other non-edible feeds, provides 6 kg (6.3 quarts) of milk for human use. This is a good investment of resources. One reason milk consumption and production is decreasing in the developed countries is the diet-health aspect related to cholesterol content and a possible link to heart disease. It is very hard to separate fact from pseudofact in this issue. Several recent studies have pointed to some very positive benefits of milk consumption. Milk is still a very important part of most major diet recommendations.

Table 3. World Milk Production

	Cow		Buffalo		Sheep		Goats		Total Milks	
	1979-81	1992-94	1979-81	1992-94	1979-81	1992-94	1979-81	1992-94	1979-81	1992-94
Africa	11,563	14,989	1,248	1,519	1,280	1,448	1,872	1,928	15,963	19,884
N.C. America	75,845	87,263	-	-	-	-	309[1]	165[1]	76,154	87,428
S. America	58,139	68,808	-	-	31	33	134	179	58,304	69,020
Asia	36,955	66,497	26,410	45,630	3,037	3,672	3,616	6,006	70,018	121,805
Europe	169,397	154,509	96[2]	138[2]	2,555	2,676	1,547	1,660	173,595	158,983
Oceania	12,248	15,991	-	-	-	-	-	-	12,248	15,991
World	420,785	460,695	27,754	47,287	7,005	7,944	7,781	10,294	463,352	526,220

[1] All from the Bahamas, Haiti and Mexico
[2] Almost all Italy and Bulgaria

Poultry and Eggs

Poultry are a source of two very valuable protein foods for human consumption, meat (discussed earlier) and eggs. They offer one of the great potentials for improving the nutritional levels in many of the developing countries. The same problem exists with poultry as with dairy cattle. The native strains are usually very hardy but are very low producers. However, when high producing strains are introduced they do not perform well because of poor feeding and management as well as disease and parasites. Despite this limitation, poultry require a low initial investment and, if necessary, can produce some food for their owners with only a minimal input in terms of feed, equipment and housing.

Eggs are a very important food source globally. Chickens produce most of the eggs for human consumption (see **Table 4**). Their production is increasing 2-4% per year on a world-wide basis. The duck, turkey, goose, guinea fowl, ostrich and some wild birds also contribute but to a much lesser degree. Eggs are an excellent source of protein (13-14%). They are also high in fat (11-15%). For poor societies this fat content is actually a positive factor because it provides calories and essential fatty acids.

Honey

Another important animal food is honey from bees. World production is estimated by the FAO at just over 1 million MT per year. However, other estimates place that much higher.

Miscellaneous Food Uses

A rather exotic way in which animals are utilized for food in some primitive areas is as a source of blood. Veins are tapped by one means or another and the blood collected. Unfortunately, this is most commonly done in areas where, because of parasitism and malnutrition, the animals can scarce afford the loss of blood. Tibetan nomads and African tribesmen (especially the Masai) do this. The fat from the tails of fat-tailed sheep is also collected from live animals in Africa and used.

Conclusion

Agricultural animals help man extend his use of the available resources by converting non-edible material to humanly edible food. This conversion is not always efficient or fast. However, allowing an animal to eat wheat straw or corn stover or to browse brush or even catch their own insects, reptiles and rodents is the only way these resources can be turned to food for humans. Man has historically consumed animal products and will go to great lengths to procure them. In so doing, man is balancing his diet which might be very low quality otherwise. The high biological value of the protein coupled with needed vitamins and minerals make animal products an essential part of the human diet.

Table 4. World Egg Production
Metric Tons

	Hen		Eggs Excluding Hen	
	1979-81	1994	1979-81	1994
Africa	945,466	1,646,930	5,214	7,929
N. C. America	5,414,971	6,349,092	490[1]	500[1]
S. America	1,569,759	2,581,442	10,783	28,765
Asia	6,962,992	18,227,216	1,145,056	2,977,055
Europe	7,220,058	6,892,153	64,413	61,157
Oceania	262,725	222,803	1,523	2,538
World	26,184,496	39,355,984	1,298,945	3,107,086

[1] All from Haiti

References

CAST. 1977. Energy Use in Agriculture - Now and For the Future. Council for Agricultural Science and Technology Rpt. No. 68.

FAO 1994. FAO Production Yearbook. FAO, Rome.

Hirst, E. 1974. Food related energy requirements. Science 184:34.

McDowell, R.E. 1977. Ruminant Products: More than Meat and Milk. Winrock Report, Sept. 1977, Winrock International Livestock Research and Training Center, Morrilton, Arkansas.

McDowell, R. E. 1991. A Partnership for Humans and Animals. Ch. 4 Foods from Animals. Kinnic Publishers. PO Box 37763, Raleigh, NC 27627-7763.

Taylor, Robert E. 1995. Scientific Farm Animal Production. 5th Edition. Prentice-Hall, Englewood Cliffs, N.J. 07632

Turk, K.L. 1975. Significance of animals in farming systems and in the provision of farm power and pleasure. Prepared for the Rockefeller Foundation Review of the Role of Animals in the Future World Food Supply, New York, April 21-22, 1975.

USDA. 1980. Foreign Agricultural Service. Foreign Agricultural Circular. FLM MT 8-80.

USDA. 1981. Foreign Agricultural Service. Foreign Agricultural Circular. FLM 7-81.

STUDY QUESTIONS

1. Compare and contrast the quantity of meat consumed by people in developed and developing countries. Use numbers.
2. What species provide the most of the world's meat?
3. Describe the rates of growth in the production of the various meats used by human societies in the world.
4. Explain how animal products in the diet help balance diets of poor quality.
5. Explain how meat production fits into the various levels of agricultural development.
6. Approximately what percent of the total milk produced in the world is produced by cattle? Which species produces the second largest amount of total milk? In what continent does the cow contribute the lowest percentage of the total milk produced, and which other species produces a large amount of the total milk in this continent?
7. What region produces the greatest percentage of the world's hen eggs. What region produces the greatest percentage of the world's eggs excluding hen eggs? Why do you suppose the developed countries don't completely overshadow the developing countries in egg production as they do in some other products?
8. Describe some miscellaneous food uses of animals.

Chapter 8

The Non-Food Uses of Agricultural Animals

Introduction

Many social, religious and economic functions are supplied by goods and services from the world's animal populations, (**Table 1**). How important each is varies with many factors including ethnicity of the owners, country and ecological condition. Non-food uses are most important in developing countries. They are much less important in developed countries but when you finish this chapter I think you'll agree that non-food uses are important to them also.

Body Coverings

For some animals, body coverings are the main product. Examples are sheep (wool, Persian lamb skins) certain breeds of goats (cashmere and mohair) and some in the llama family. For other species, the body covering is a by-product rather than the main product. A major advantage of the fiber products is that they are renewable. This means they can be harvested repeatedly from the same animal. Skins and hides are non-renewable, at least from the same animal. In countries with a highly developed agriculture, the skins and hides obtained at time of slaughter are a valuable by-product. However, their contribution to the total value of the slaughter is relatively small compared to the value of the carcass. Often, they are shipped to a developing country for processing. In contrast, in some areas of low subsistence or primitive agriculture, hides or skins rather than the meat may be the most valuable animal product that is exported from that country and may provide valuable foreign currency.

World hide production is given in **Table 2.** India is the largest producer of cattle and buffalo hides with the USA a close second. Argentina, Brazil, Germany and China all produce large quantities as well. Most sheepskins are produced in Asia with Bangladesh and China producing the most. Asia also produces the most goatskins with China, India and Pakistan being the individual countries producing the most. We should point out that hides and skins (roasted pig especially) are used for food in many countries and cattle hides are eaten in Africa and a few other places.

Wool production is given in **Table 2**. Oceania dominates world wool production, producing almost 40% of the world's total. Australia is the most important producer with New Zealand contributing its fair share, especially considering the small land mass of New Zealand. China is also a large wool producer.

Goats yield 2 types of body coverings that are used for making fabrics - mohair and cashmere. Mohair is clipped from the Angora goat each year in the same manner as the fleece is shorn from sheep. The average yield is about five pounds. Angora goats are raised in several parts of the world. Cashmere is the very fine underfur of the Kasmir goat. It must be combed out of the

Table 1. Some contributions of Services and Non-Food Products

Classification	Contribution	Main Sources
Traction	Agriculture	7 species ruminants, 5 species non-ruminants
	Cartage	Cattle, buffalo, yak, camel, horse, mule, donkey
	Packing	Camel, yak, horse, mule donkey, reindeer
	Herding	Horse, mule, camel
	Irrigation pumping	Buffalo, cattle, camel
	Threshing grains	Cattle, horse
	Passenger conveyance	Horse, donkey, mule, buffalo, camel
Storage in Animals	Capital	All domestic species
	Grains	Buffalo, cattle, sheep, pigs, fowl
Conservation	Grazing	All domestic herbivore
	Seed distribution	All domestic herbivore
	Soil through crop rotation	Most grazing animals
Ecological	Maintenance	Most animals
	Restoration	Most animals
Pest control	Weeds between crops	Domestic ruminants, ducks, geese
	Insects between croppings	Poultry, ducks, geese
	Irrigation canals	Buffalo, ducks
Cultural needs	Exhibitions including rodeo	Horse, cattle, goat sheep, pig, buffalo, poultry
	Bride price	Cattle, goats, sheep
	Religious sacrificial (Muslim)	sheep, goats
	Fighting	Cattle, poultry, buffalo,

		sheep
	Hunting	Most all species
	Pet	Numerous species
	Racing	Horse, cattle, dogs
	Status symbol	Horse, cattle, buffalo
Capital generation	Security	Cattle, buffalo, goat, sheep, pig
	Liquidity	Most all species
	Reduce risks of cropping	All domestic species
Labor	Output of cropping	Traction species
	Leisure time	Traction species
Non-arable lands	Income	Cattle, camels, goats, sheep
	Soil fertility in cropping	Cattle, goats, sheep, buffalo
Research	Numerous	Domestic & Laboratory
Fiber	Wool, Hair, Feathers	Sheep, camels, llamas, alpacas, yak, horse and other mammals. All Fowl
Skins	Hides, Pelts	Most all species
Waste	Fertilizer, fuel, methane gas, construction, feed (recycled)	Most all species
Inedible Products	Fat, horns, hooves, bones, tankage, endocrine extracts	All species
Income		All species

Source: Adapted from McDowell, p 17 and 45 1991.

Table 2. World Hide and Fiber Production
Continents and Selected Countries

	Silk, MT		Wool, Greasy, MT		Cattle and Buffalo Hides, MT		Sheepskins, Fresh, MT		Goatskins, Fresh, MT	
	1979-81	1994	1979-81	1994	1979-81	1994	1979-81	1994	1979-81	1994
Africa	22	21	2,785,614	2,665,751	453,625	539,650	132,226	160,113	84,992	114,628
N. C. America	-	-	56,670	37,321	1,335,168	1,191,630	22,876	23,706	9,938[3]	9,599[3]
S. America	1,300[1]	2,450[1]	298,272	241,920	964,660	1,103,768	75,850	72,282	11,146	13,702
Asia	62,640	111,857	441,714	538,868	1,235,333	1,873,766	604,332	927,423	233,750	460,196
Europe	435[2]	181[2]	276,213	248,273	1,225,407	1,090,186	187,946	220,077	15,024	17,107
Oceania	-	-	1,057,405	1,013,000	239,234	285,121	228,517	230,332	1,630	6,162
USA	-	-	48,863	31,000	990,501	862,033	17,128	15,000	-	-
China	34,199	84,001	172,667	260,000	83,603	383,926	62,664	186,648	42,591	132,927
India	4,926	17,000	-	-	735,999	924,094	36,660	39,780	75,120	112,702
Japan	15,642	2,400	-	-	-	-	-	-	-	-
Australia	-	-	704,638	735,000	193,000	228,100	118,842	130,325	1,414	5,615
New Zealand	-	-	352,767	278,000	43,700	53,500	109,667	100,000	105	368
Argentina	-	-	-	-	437,060	366,000	38,209	22,750	-	-
Brazil	1,300	2,450	-	-	284,133	432,000	-	-	4,000	6,100
Uruguay	-	-	-	-	-	-	13,589	23,905	-	-
Bangladesh	-	-	-	-	32,553	32,543	306,133	443,100	12,425	30,745
Pakistan	-	-	-	-	62,717	123,398	33,933	57,200	46,491	107,980
Turkey	-	-	-	-	-	-	54,928	64,400	-	-
Germany	-	-	-	-	250,500	198,000	-	-	-	-
France	-	-	-	-	170,333	163,000	21,833	16,200	-	-
Russian Fed.	-	-	-	-	320,000	315,000	42,000	38,000	-	-
World	68,022	114,509	2,785,614	2,665,751	6,163,861	6,282,676	1,377,456	1,647,174	362,088	623,031

[1] All Brazil
[2] Mostly Bulgaria and Romania
[3] Almost all Mexico.
Source: FAO. 1994 Production Yearbook.

coarse outer guard hair by hand. The average yield is only about 3 ounces. Kasmir goats are raised in the high plateaus of Tibet and northern India.

Yaks produce a coarse fiber from which clothing, blankets and one of the best horse blankets known to man are made. The long hair of the camel is woven into cloth or blankets primarily for the use of the people owning the camels. Camel-owning nomads also use camel hair as tent fabric material. Camel hair brushes are highly valued by some artists. The other animals of the camel family yielding fiber are the llamoids found in the Andes mountains of South America, the llama, alpaca and vicuna. The domestic llama is actually a composite of four previously wild strains. The alpaca produces a wool that is finer than most sheep. The vicuna is a wild species of llama that produces the finest wool known. These animals are hunted for their wool which is very expensive and in demand. Luckily a large herd is protected.

The Karakul and Kuche are furbearing breeds of sheep. The lambs are slaughtered within a few days after birth and the pelts removed. The Karakul is the best known of the two, yielding Persian Lamb Skins used in the manufacture of very expensive coats. The Karakul thrives in very dry environments. Most are produced in Afghanistan and in Nambia, Africa. The skins are exported to the developed countries. There have been a few unsuccessful attempts to produce them commercially in the USA. Generally, these efforts have failed because they could not compete costwise with cheaper land and labor of Africa and Asia. The pelt of the Kuche is used in making fur hats, collars, and coats. It is seldom exported out of northern Asia.

Feathers are used for pillows and bedding, clothing (down coats) and for jewelry and adornments on clothing (hats), fishing lures and as a protein supplement to feed animals. Silk (**Table 2**) from silkworms is important in production agriculture in Asia (China, India, Korea and Japan) and in Brazil. "Wool, hair and feathers are used in cottage industries in over 100 countries to produce clothing, bedding, carpets and handicrafts. Half of the people of southwest Africa, the Andes mountains of South America and the Himalayan mountains of Asia are dependent on sales of animal products and bird feathers as handicrafts."

As a Source of Power

In a developed agriculture, almost all farm power is provided by machines rather than by animals. In developing countries, however, nearly 85 percent of the total non-human power used in crop production and agriculture throughout the world is still provided by animals. Africa, the Near East, Latin America and the Far East still use 95% or greater animal power. Countries most dependent on animal power are given in **Table 3.** For the purposes of this discussion, draft power refers to plowing, cartage (e.g. pulling a sled) and carrying loads on or over the back (packing). Oxen (steers of cattle), donkeys, horses, buffaloes, mules and camels are the major species used. There is some use of several other species in specific geographic location such as yaks, llamas, goats, reindeer and dogs. Oxen and buffaloes are the most powerful in pulling with oxen being faster on dry land. Horses move the fastest which is an advantage if high quality implements are available. Donkeys, camels and yaks can carry the heaviest loads in relation to body size and are preferred in drier climates.

Table 3. Working Animal Populations (Million) in 18 Countries.

Country	Cattle	Buffalo	Camelids
India	110	162	
China	53	17	1
Bangladesh	10	1	
Thailand	3	5	
Pakistan	7	1	1
Ethiopia	6		1
Indonesia	4	2	
Burma	4	1	
Nepal	3	2	
Philippines	1	3	
Mexico	3		
Brazil	3		
Turkey	3		
Colombia	1		
Tanzania	1		
Egypt	1	2	
Chile	1		
Peru	1		1
Total	215	196	4

Source: McDowell, 1991. P 49.

For most of the developing world there is really no alternative to draft power. "Globally there are 25 million animal-drawn vehicles (Faulvey 1989). Animals generate 150 million horsepower. Their estimated market value is $100 billion. Replacement of these by mechanization would cost over $300 billion of capital, $10 billion per year for fuel and $20 billion for repairs."

Miscellaneous Uses of Animals

The following miscellaneous uses of animals are also important. (McDowell 1977, McDowell 1991.) We first discussed these in the Introduction and now will discuss them individually.

- Wastes for fertilizer, fuel, construction material, and pest control.
- Pest and weed control.
- Storage of capital and food.
- For cultural and religious reasons and as a status symbol. Recreation, exhibitions (including rodeos), fighting, riding and pets.
- Nonfood and inedible slaughter by products.
- Income.
- Conservation.
- Human Health Research

Body Wastes

Manure. As many as half of the world's farmers may depend on animal manure and compost from plant residues as their only source of fertilizer. In fact it is preferred over chemical fertilizers by small farmers who till the land by hand or with crude plows because manure improves soil texture. Manure is also used in the developed countries as a source of fertility, but not very efficiently. Domestic animals in the US produce 2 billion tons of manure per year, with 50% collectible. If all of the collectible manure was spread on cultivated land, it would supply 19% of N, 38% of P, and 61% of the K in the chemical fertilizers used each year. (Ensminger, 1991). There are many of the poorer areas of the world where manure must be utilized as a fuel. This is particularly true in very densely populated areas where the trees have long since been cut and used as fuel or for other purposes. In these areas, dried manure, particularly from cattle and buffalo, is an important source of fuel for fires for cooking and heating (200+ million tons each year). India, Central Africa, the Andes of South America, the highlands of Central America and Caribbean Islands all use a great deal of manure for fuel. The sale of this product is an important source of income for many farm families who take dried dung cakes to the cities and sell them. A third use of manure is in construction. Many in Africa and Asia use cow and/or buffalo manure as plaster in their huts. In the upper elevations of Peru and Uruguay, houses are built of blocks made from a mixture of approximately 50% manure plus soil and straw. In other cultures, cow manure is used as an important ingredient in poultices for wound healing (possible antibody effect).

Urine. Urine also has some specialized uses. In Asia cow urine is sprinkled on the dirt floors to control dust and pests. In Africa urine and blood are used to extend the low volume of milk available for human food. It is even used as a rinse for hair.

Pest and Weed Control

Land that is being left to fallow (rest) can be grazed by animals to control weeds and add their manure back to the soil to rejuvenate it. Grazing animals are also frequently used along the edges of canals to control vegetation. In addition to weed control the animal is supported. Grazing the canals also helps control snail-borne diseases in the human population because large numbers of snails are killed under the hooves of the grazing animals. Ducks are also used in irrigation canals because they eat snails and leaches. It is common to use ducks in rice fields after harvest to pick up missed grain and to harvest insects. Animals provide a tremendous service globally by reducing weeds and crop residues on cropland. These must be removed from land if it is to be prepared by hand tools or animal drawn plows. One method used to accomplish this is to heavily stock the land with animals to remove the weeds, grasses and residues just before planting. Farmers and pastoralists in Africa often have a centuries-old arrangement to provide mutual service to each other in this way. Consumption of these nuisances to cropping by animals adds additional economic value to the cropland through the animal product gained. In Africa, close grazing of grasses and weeds adjacent to cropped areas is used to minimize the presence of insects by creating a vegetation free barrier around the crop. The use of animals to break cycles of insects and assist in weed control could have potential for the developed world and allow reductions in the use of pesticides in certain areas.

Storage of Capital and Foods

In many of the poorer developing countries banks are nonexistent. Even if they exist, the poorer people either do not have access to them or do not trust them. As an alternative to banks, people use livestock (particularly cattle) as a vehicle for storing their surplus capital. If money is needed, animals can be sold or often bartered for other needed items. Use of livestock for storage of capital has several undesirable effects. Obviously, large herds are desirable and are acquired and maintained. This often results in severe overgrazing. This is a major reason why desertification increases when the human population increases. A second very undesirable effect is the reluctance of people to sell animals unless money is needed. This prevents the development of an orderly program of marketing and slaughtering of animals such as is found in the developed countries. For maximum utilization of the livestock of a country there must be such a marketing program in place in the country.

The use of animals as storage of food is very important in many countries of the world since they represent one of their most reliable reserve food supplies. Most poor countries have inadequate facilities for storing grain (especially for any long period) without suffering large losses from pests and storage. This is not a problem with animals. Animals assume even greater importance as a storage of food in countries subjected to periodic and prolonged droughts. When the rains do not come, grains cannot be planted or, if planted, die. Animals can continue to survive for several months, even years, and may be the only means of survival of the human population.

Developed countries do this also. It is estimated that the US and developed countries have enough food stored in their livestock to feed their populations for 1 year.

Cultural

Both domestic and wild species of animals provide for the cultural needs of human societies. The roles of animals in food supplies and for services can be secondary to these cultural roles. Some of these are listed below. Religion is a very important cultural role and was discussed earlier.

Exhibitions. Animals are exhibited for a wide variety of reasons and in almost all countries. The reason can be as simple as advertisement by breeders interested in selling breeding stock to others. However, the love of competition is an important motivator as well. Some believe the purebred livestock industry in the US would disappear without this social dimension. The excitement of this activity is catching and is used to motivate interest by young people in animals. There is a huge 4-H and FFA Club youth project in the USA. If we add such activities as rodeos, fighting (bulls, buffaloes, sheep and chickens), racing (camel, horse, buffalo, dog, fowl, turtles, frogs and horses), then together they provide around 10 billion person days of recreation annually. In developing countries, these combined animal recreation events are by far the largest source of entertainment for people.

Pets. Many species are used as pets and companion animals. New species, such as the alpaca, llama, small pigs, hedge hogs and others are becoming used as companion animals. Domestic livestock are also used extensively as pet species in both developed and developing countries. There was a time when the human animal bond was viewed as nothing more than a sentimental gesture on the part of humans to "dumb" animals. However, modern research has demonstrated a myriad of benefits to humans from pet ownership.

Social structure. The more primitive the society the more livestock, particularly cattle, are a part of the cultural fabric of the people. Animals are often the traditional legacy a man leaves to his sons since they are the only significant possessions a man can acquire. This is especially true of nomads. They are a traditional way that formal contracts are sealed. Marriage contracts involve exchanges of livestock. In tribes where polygamy is practiced, there is the pressure to accumulate animals in order to obtain additional wives. In some tribes the traditional bride price is so high that young men must work for several years to obtain the necessary livestock. This thus serves as a means of birth control.

This has been only a partial listing of the contributions animals make to the culture of a people. It is important to remember that a very high priority is given to this by the world's people. In some cases the most important reason for keeping animals by a given group of people is related to the role animals play in their culture.

Non-Food or Inedible Slaughter By-Products

Many things that developed countries consider inedible are eaten in developing countries. However, there are a large number of non-food products derived from slaughter at all levels of development. A partial listing of sources and products follow:

Source	Some Products
Blood	Glue, Feed, Vaccine, Serum
Bones	Buttons, Jewelry
Digestive Tract	Feed, Fertilizer
Endocrine Glands	Pharmaceuticals
Inedible fat	Fatty acids, Fuel, Soap, Candles
Hooves	Gelatin, Glue
Horns	Handicrafts, Cultural Symbols

Source: McDowell, 1991, p. 42)

In developed countries, the largest product is inedible fat, used widely for industrial purposes. Of equal monetary value, are the large number of products used in medicine. A partial listing includes epinephrine, thrombin, insulin, liver extracts; cortisone, norepinephrine, plasma, blood fibrin, heart valves, estrogen, relaxin, pepsin, oxytocin and burn dressings. Life preserving vaccines are produced on animal tissue and on chicken eggs. For example, the measles vaccine is produced on canine kidney as well as on eggs.

Income

With the exception of pets, the reason developed countries have livestock industries is to make money. In the U.S. annual income from animals approaches $90 billion, rising on average at 3-4% yearly (1996 numbers). Labor generated by the animal industries is one of the most significant employment sources. Animal products rank first or second as a source of income from agriculture in over 40 of the 50 states. Developing-country agriculture also depends on animals for income. The primary argument we use to justify greater livestock production in developing countries is simply to feed people better. A more important reason is the contribution they make to total farm production and economic growth. This is so because they are the primary source of cash income for nearly all farms. Small increases in animal production quickly and substantially increase farm income making available cash to purchase fertilizer and other items which contribute to better production. Increases in overall food supplies, lower food costs and increased employment opportunities combine to cause an improvement in the diets of the people. Overall development of the economy occurs through cropping because the animals aid soil cultivation and fertility of the soil and through demand for feed supplies, processing products and marketing services. In rural areas where employment opportunities are low, the use of animals provides a stable labor demand throughout the year. They are also a highly liquid asset for emergency use. Withdrawal of these resources would be devastating and would add immensely to the already impoverished situation in many countries.

Conservation

Contrary to common belief and media hype, domestic animals play important roles in soil and environmental conservation. They are a vital tool in many land conservation practices. These include:

- Controlling pests and adding manure to lands being left to fallow.
- Alley cropping systems where leguminous trees are planted in strips with animals and crops alternated between them.
- Providing alternative use in tree cropping systems and "weeding" stands of palms.

Human Health Research

The number of applications here is simply too numerous to mention. It is safe to say that 20-25 years have been added to the average American life span due to health research conducted on animals. In addition, the quality of life for people afflicted with chronic diseases has been helped immeasurably. This is an area especially contentious to animal rightists. Ingrid Newkirk of PETA says, "A rat, is a dog, is a pig, is a boy." Luckily, most people are more rational. The simple truth is that medical research depends on the use of animals as models. It will continue to do so for the foreseeable future. The choice is simple, if you care about people.

Conclusion

Animals provide man with a myriad of services in addition to food. In this chapter there were many discussed. There are many more not discussed. Suffice it to say that without the goods and services provided by animals in addition to food, peoples in developed and developing countries would find it difficult to maintain their living standard, perhaps even survive.

References

CAST. 1977. Energy Use in Agriculture-Now and For the Future. Council for Agricultural Science and Technology Rpt. No. 68.

Ensminger, M.E. 1991. Animal Science (Animal Agriculture Series). 9th edition. Interstate Publishers, Inc., Danville, IL 61834-0050.

FAO. 1984. FAO Production Yearbook. FAO, Rome.

Hirst, E. 1974. Food related energy requirements. Science 184:34.

McDowell, R.E. 1977. Ruminant Products: More than Meat and Milk. Winrock Report, Sept. 1977, Winrock International Livestock Research and Training Center, Morrilton, Arkansas.

McDowell R. E. 1991. A Partnership for Humans and Animals. Ch. 5. Non-Food Products. Kinnic Publishers. P. O. Box 37763. Raleigh, NC 27627-7763. Author's note. This chapter borrows very heavily on Dr. McDowell's work and is done so with his permission. The text cited here is an excellent book. However, Dr. McDowell has no intentions of revising and updating the text. Therefore, he has graciously granted consent for extensive use of the material.

Taylor, Robert E. 1995. Scientific Farm Animal Production. 8th edition. Prentice-Hall.

Turk, K.L. 1975. Significance of animals in farming systems and in the provisions of farm power and pleasure. Prepared for the Rockefeller Foundation Review of the Role of Animals in the Future World Food Supply, New York, April 21-22, 1975.
USDA. 1980. Foreign Agricultural Service. Foreign Agricultural Circular. FLM MT 8-80.
USDA. 1981. Foreign Agricultural Service. Foreign Agricultural Circular. FLM MR 7-81.

Study Questions

1. What are the various body coverings provided by animals? Which animal produces which product?
2. What is meant when we say that skins and hides are non-renewable but wool is renewable?
3. What is different about the Karakul and the Kuche breeds of sheep?
4. What is the value of animal fibers, feathers, etc., in the cottage industries of poor countries?
5. What are the various uses which we call draft power?
6. What is the estimated market value of the draft animals of the world?
7. List and briefly describe the miscellaneous uses man has for animals.
8. Name 10 non-food slaughter by-products man gets from animals.
9. How do animals aid in land conservation?

Chapter 9

Cattle

Introduction

Cattle are clearly among the most useful of all the domestic species and are used by a wide variety of the world's peoples for a large number of purposes. It is arguable that they have been the most useful of all domesticated farm animals. They are certainly the most wide spread and numerous of all the species. In this chapter we will explore their contributions to man.

Classification

Cattle belong to the genus **Bos** with two species; **taurus** (European breeds) and **indicus** (Zebu cattle) which have humped necks. The two species interbreed and produce a fertile hybrid. The two species of cattle are fairly clearly segregated into distinct portions of the world. Taurus is generally the choice in temperate regions. They predominate in North America, Europe, Australia, and southern South America. Indicus is the predominate species used in unfavorable environments, particularly hot climates. This includes southern Asia, most of Africa, northern South America, Central America and southern Mexico. Taurus is usually more productive than B. indicus and are usually raised if at all possible. This is especially true if the intended use is milk or meat.

Uses of Cattle in a Developed Agriculture

The level of agricultural development dictates the uses the population has for cattle and therefore the type of cattle that will or will not be present. As has been discussed at various places in this text, the more highly developed the agriculture the more likely livestock production is to be highly specialized and the animals to be kept for the production of one product because profit is easier to make this way. In the case of cattle, the product is either meat or milk. Developed agriculture keeps farm herds that are either beef or dairy cattle and range herds that are always beef and never milk animals.

Dairy Herds. The dairy industries of all the developed countries are very similar. They are almost always based on the Holstein if adapted. They are always intensively managed. A highly developed US-style dairy industry, exists only in countries with a developed agriculture. Generally, efforts to establish a similar industry in a developing agriculture have failed. When such an attempt is made trying to use the local strains of cattle, it fails because of the poor genetics for milk yield of the native cattle. Attempts to introduce breeds with the genetic potential for high milk production usually also fail for two important reasons. Probably the most important reason is that the people have never fed supplemental feed to cattle and will not do so to the introduced breeds. As a result the introduced cattle produce at only a fraction of their potential and not much higher, if any, than the native cattle. The other reason is the generally unfavorable physical environment (climate, disease and parasites) that is typical of many developing countries. Success would require not only better feeding but also better health management than the people know how to provide or are willing to provide.

Beef Cows. First world countries having major range beef operations are the United States, Australia, Canada, and South Africa. Lesser developed and developing countries having major range beef operations are Argentina, Mexico, Uruguay and Brazil. Range herds are extensively managed in order to reduce labor costs and increase profits. Most of Europe keeps beef cattle in farm herds. Sometimes these cows will be milked for home consumption especially in the parts of Europe that are on the fringes of being developed. Dual-purpose breeds are often kept in these herds. Dairy herds also produce meat. Cows are used for meat in salvage situations. Bull calves and extra heifers are used for meat. The primary purpose of these herds is milk and they are classified as dairy because of this.

Uses of Cattle in Subsistence Agriculture

As we have stated before, in subsistence agriculture the farm must produce most of the needs of the family. Animals are the primary source of power and cattle are the animal of choice for power except in rice paddies. Cattle also supply milk, meat, leather, manure and other products. Strains of animals best suited to subsistence are much less productive in meat and milk than "improved" breeds but do a better job meeting the total needs of subsistence agriculture. Triple purpose animals predominate but the more developed the agriculture the more dual-purpose animals are found. As an agriculture reaches high subsistence, the farmers may come to depend on horses and mules for work animals because they can move faster. High subsistence farmers can afford higher quality plows.

Uses of Cattle in Primitive Agriculture

In many primitive agricultures, non-food uses for cattle are more important than food. Use of animals as meat will be in a salvage situation almost exclusively but milk and blood will be used. There are several non-food uses for agricultural animals for which cattle are preferred. Cattle are preferred over all other animals for storage of capital and for cultural reasons. They serve as a status symbol, indicator of wealth, as "capital" in trade and contracts, and the major form of inheritance from father to son. If they predominate, then one should also expect them to do work (if any animal power is used), provide for transportation, and supply manure for the fields. However, most primitive agriculture does not use much animal power but rather relies on hand labor. Nomadic societies often use cattle as one of their species if it is adapted to the particular climate.

Distribution and Use of Cattle in the World

The leading countries of the world in numbers of cattle and the production of beef and veal are presented in **Table 1**. India, where cattle are sacred, has the largest population of cattle of any country in the world. However, since cattle cannot be slaughtered or sold for slaughter by the Hindus in India, they yield little fresh meat in comparison to their numbers. The slaughter there was about 12.5 million head in 1994. The slaughter was accomplished by the non-Hindus which

Table 1
World Cattle Numbers and Meat Production

Rank	Country	No. Cattle (1,000 head)	Meat Rank	Meat Production 1000 MT
1	India	192,980	11	1,292
2	Brazil	151,600	3	3,160
3	United States	100,988	1	11,199
4	Russian Federation	48,900	2	3,350
5	China	90,906	5	2,253
5	Argentina	50,000	4	2,590
7	Mexico	30,702	10	1,365
8	Ethiopia	29,450	*	230
9	Columbia	25,700	*	609
10	Australia	24,732	6	1,825
11	Bangladesh	24,130	*	145
12	Ukraine	21,607	9	1,421
13	France	20,112	7	1,667
14	Pakistan	18,146	*	338
15	Germany	15,891	8	1,500
16	Venezuela	15,071	*	370
17	Tanzania	13,376	*	202
1	Asia	410,118	4	6,619
2	S. America	279,656	3	7,946
3	Africa	192,180	5	3,264
4	N.C. America	163,354	1	14,086
5	Europe	107,158	2	9,564
6	Oceania	34,049	6	2,414
	World Total	1,288,124	**World Total**	50,509

[a]Carcass weight.
[b]Sources: 1994 FAO Production Yearbook.
*Rank for meat production not provided.

comprise approximately 10% of the population. The top countries in meat production from beef are either Developed Countries or large countries. USA, Argentina, Australia, Mexico and Brazil all have well-developed range industries. Only one of the top 5 countries is an Asian country (China) and the only African country in the top 10 for animal numbers is Ethiopia. Cattle produced about 25% of the total world's meat production.

It is important to notice the difference between the number of animals and the production of meat from them. For instance, the U.S. is ranked third in number of animals but first in production of meat. India is first in numbers of animals but ranks very low in meat produced. The U.S. has only about one-fourth the cattle of all of Asia but produces 1.7 times as much meat from them. The differences are even more staggering if one considers that the population of Asia is almost 13 times larger than that of the U.S. A similar situation can be observed if the U.S. is compared to all of Africa. There are also tremendous differences in the rapidly developing countries of Latin America and less developed Asian and African nations. Mexico and Ethiopia have very similar cattle populations but Mexico produces almost six times as much meat. A similar situation is evident between Pakistan and France. It is important to remember that this low meat production by the poorer nations does not mean that they are not deriving benefit from their cattle. It just means that meat from cattle is much less important than other needs and reflects the difference in emphasis of First World versus Third World Countries. In Asia and Africa, non-food uses are frequently more important than meat production.

Milk

The leading countries in the production of cows milk and dairy cow numbers are presented in **Table 2.** India is the largest producer of water buffalo milk and goats milk of all the world's countries and ranks third in cow's milk production. Milk is very important in a Hindu society. Notice that China ranks low in numbers of dairy cows, especially in relation to their human population. They certainly have the opportunity to produce more. The reason they don't produce more can be found in our earlier discussion regarding religion. Do you remember? At any rate, modern Chinese are not as a whole especially fond of dairy products. Africa and Asia combined only produce around 17% of the world's cows' milk, in spite of their dense population. A look at the production per cow is also instructive. Those with the highest productions are either developed or have a developed dairy industry within a lower level agriculture. The average cow in Israel gives 21 times as much milk as the average African cow. The USA has the highest total fluid milk production and Israel produces the most milk per cow (9389 Kg). The world's average dairy cow produces 2,034 lbs. of milk per year.

World butter and ghee production is 6,711,631 MT. The United States, India, the Russian Federation, France, Germany, Pakistan and the Ukraine are the leading countries in butter and ghee production. The United States, France, and Germany are the top cheese producers with all countries producing 14,880,089 MT of cheese.

Asia and Africa-General Considerations

Table 2
World Dairy Cattle Numbers and Milk Production

Rank	Country	Total Dairy Cattle (1,000 head)	Milk Production Per Cow/Kg	Total Milk Production (1,000 MT)
1	India	30,500	984	30,000
2	Brazil	20,000	789	15,774
3	Russian Federation	20,000	2,200	44,000
4	United States	9,575	7,277	69,682
5	Ukraine	8,003	2,241	17,933
6	Mexico	6,480	1,165	7,547
7	Turkey	6,135	1,459	8,950
8	Sudan	5,441	480	2,610
9	Germany	5,301	5,320	2,820
10	Colombia	4,700	998	4,690
11	France	4,692	5,314	24,935
12	Pakistan	4,600	891	4,100
13T	Kenya	3,866	493	1,905
13T	Poland	3,866	3,161	12,219
15	Bangladesh	3,761	206	774
16	China	3,625	1,545	5,600
1	Asia	59,129	1,135	67,129
2	Europe	35,723	4,270	152,550
3	Africa	34,057	446	15,197
4	S. America	32,204	1,117	35,958
5	N. C. America	20,116	4,389	88,297
6	Oceania	4,675	3,590	16,783
	World Total	225,498		485,645

Sources: 1994 FAO *Production Yearbook.*

Asia and Africa have most of the very poor nations of the world. With the exception of Israel, Japan and South Africa, which are classified as First World, all of the nations are classified as developing countries. Therefore, it should be expected that most of the agriculture is subsistence or lower and the most important use made of cattle in many countries is nonfood. In addition, a large part of these two continents has less than a desirable environment in terms of climate, diseases, insects and parasites.

Cattle in Asia. At least 45 important breeds or strains of cattle have been developed in Asia. Most do not have a breed association or a herdbook, therefore, by standards of the First World they are not breeds but strains. However, many of them have all of the attributes of a breed. Most are Zebu's, which is Asia's most important contribution to the cattle of the world. Most were developed in a specific section of India and, except for a few of the most outstanding strains, are still primarily located in these areas. Their name was usually derived from the name of the area. The breeds of Zebu that have been exported to other parts of the world have largely gone to areas with unfavorable environments, usually hot and humid.

Cattle in Africa. The types of African cattle may be divided into three groups. These include unimproved, native strains consisting of both humped and non-humped cattle but the majority are humped. The term sanga has come to be used to refer to these native strains as a whole. The rest are improved native strains, the most important of which are the Africander and Boran and exotic breeds that are primarily Zebus from Asia.

Conclusion

Cattle have proven to be among the most adaptable and useful of all the world's agricultural animals. They have thus become the most widely distributed of all agricultural animals. Both highly developed and primitive agricultures make use of them. Developed agricultures have very specialized uses while poorer societies find more general uses for them. There is also a near universal prestige that cattle ownership conveys that transcends economic boundaries.

References

Briggs, H.M. and D.M. Briggs. 1980. Modern Breeds of Livestock, 4th Ed., MacMillian. New York.

FAO 1994. Production Yearbook. FAO, Rome.

French, M.H. 1966. European Breeds of Cattle. Vols. 1 and 2. FAO Agricultural Studies No. 67. FAO. Rome.

Genetics and Economic Implication. Texas Agr. Exp. Sta. Tech. Rpt. No. 12.

Taylor, R.E. 1995. Scientific Farm Animal Production, 5th ed. Chapter 2.Prentice-Hall, Inc., New Jersey.

USDA. 1976. Foreign Agriculture Service Circular, FW 1-76.

Study Questions

1. What are the genus and species of cattle?
2. Describe how cattle are most commonly used in a developed agriculture. Subsistence agriculture. Primitive agriculture.
3. What are the major cattle owning countries? What are the major meat producing countries?
4. What are the major cows' milk producing countries?
5. What is unique about cattle in Asia?
6. What is unique about the cattle in Africa?

Chapter 10

Sheep and Goats

Introduction

Sheep and goats are important species to the world's peoples. Both are ruminants and are found in temperate and tropical regions and in primitive, subsistence and developed agriculture. They provide hair, wool, meat, milk and hides. Sheep and goats have better water and nitrogen economy than cattle. That makes them better adapted than cattle to the arid tropics. There are more sheep in temperate zones and more goats in the tropics. Temperature and type of vegetation found in the respective zones are the principal reasons for this segregation. Sheep are much more productive in temperate zones than tropical areas. There are several reasons for this. First is the more favorable climate followed by a better feed supply. In addition, since most developed agriculture is in the temperate zones this means that the animals there have been genetically selected to be better in the productive traits. Unfortunately, sheep from temperate zones are so well adapted that they do not do well when taken to a tropical zone and thus are little help in upgrading these animals.

World sheep numbers (see **Table 1**) demonstrated some modest growth in the 1980's, but have declined to about 1.0 billion head in 1994. This is approximately the same number of sheep there were in 1980. Numbers are increasing in Asia, probably as a food species for the growing population, declining in South America as producers turn to cattle, declining in the former East Block countries and in the Russian Federation because rough economic times are causing the population to eat the breeding flocks, declining in Oceania due to drought in Australia and increasing modestly in North America.

The 609 million goats in the world can be largely accounted for by the few countries listed in **Table 2** . The top 5 goat producing countries have over 50% of the total goat population. Goats are important for both milk and meat production. Goat numbers have been increasing for the last 15 years or so in response to growing populations in the countries where it is found in largest numbers.

Sheep

There is a wide variety of sheep in the world. In many areas of the world they have hair rather than wool. This hair may be fine and thick, producing a pelt as in the Karakul breed which yields Persian Lamb Skins. It is usually a coarse hair with no value as a fiber. Sheep with a covering of coarse hair are usually found only in hot areas with low subsistence to primitive agriculture. In these areas the people primarily keep animals to produce for their own needs and have no need for wool because of the hot climate. They do need meat and milk. The utilization of sheep in different economic situations is summarized in **Table 3**. The number opposite a use ranks that use as to its relative importance.

Table 1. World Sheep Numbers, Meat Production and Greasy Wool Production.

Country	Numbers (X 1,000)	Word Rank	Meat Production 1,000 MT	World Rank	Greasy Wool Production MT	World Rank
Australia	132,609	1	642	3	735,000	1
China	112,649	2	800	2	260,000	3
New Zealand	50,135	3	482	4	278,000	2
Iran	45,400	4	254	7	34,500	
India	44,809	5	171		35,000	
Russian Fed.	41,078	6	850	1	145,000	4
Turkey	37,541	7	300	5	41,000	
Kazakhstan	33,524	8	220	10	100,100	5
UK	29,300	9	278	6	65,888	9
South Africa	29,134	10	136		77,763	8
Pakistan	28,975	11	246	8	52,000	10
Spain	23,838	12	228	9	29,000	
Uruguay	23,441	13	75		78,000	7
Ethiopia	21,700	14	77		11,500	
Sudan	22,870	15	75		20,600	
Brazil	20,500	16	84		30,750	
Argentina	20,000	17	70		92,000	6
Algeria	17,850	18	142		50,100	
Mongolia	14,392	19	95		18,093	
Afghanistan	14,200	20	115		15,000	
Romania	11,499	21	72		28,220	
France	10,452	22	140		22,000	
USA	9,600	23	151		31,000	
Asia	340,102	1	2,342	1	538,868	2
Africa	208,845	2	890	4	229,325	5
Oceania	182,758	3	1,124	3	1,013,000	1
Europe	130,692	4	1,242	2	248,273	3
S. America	94,054	5	292	5	241,920	4
NC America	17,529	6	199	6	37,321	6
World	1,086,661	-	6,886	-	2,665,751	-

Table 2. World Goat Numbers, Meat Production, and Goat Milk Production.

Country	Number x 1000	Rank	Meat Production (1,000 MT)	Rank	Milk Production (1,000 MT)	Rank
India	118,347	1	470	2	2,351	1
China	105,990	2	751	1	181	5
Pakistan	41,340	3	391	3	620	3
Bangladesh	28,050	4	98	5	1,048	2
Nigeria	25,497	5	130	4	?	
Ethiopia	16,700	6	62		93	
Sudan	16,449	7	37		590	4
Brazil	12,200	8	35		11	
Mexico	10,450	9	39		141	
S. America	22,819	3	74	4	182	4
N. C. America	14,944	4	49	5	159	5
Europe	14,809	5	90	3	1,633	3
Oceania	1,007	6	22	6	--	-
Asia	373,005	1	2,135	1	6,135	1
Africa	176,089	2	662	2	2,003	2
World Total	609,488		3,057		10,480	

1994 FAO Production Yearbook

Table 3. Summary of the Major Utilizations Made of Sheep in Various Levels of Agricultural Development.

Level of Agricultural Development				
Developed			Subsistence	Primitive
Predominate Production System				
Intensive	Extensive			
Farm Flocks	Wet Range	Dry Range		
1. Meat 2. Wool	1. Dual Purpose	1. Fine Wool 2. Meat	1. Triple Purpose 2. Pelts	1. Meat 2. Milk 3. Skins, Wool

The economics of producing both cattle and sheep in the developed countries are quite similar. These two species are usually the ones competing for the use of grasslands of a developed country. The single most important thing determining where there will be large concentrations of either cattle or sheep is the availability of relatively cheap roughage. When both are equally well adapted to a region, cattle are usually the species of choice. However, there are a few exceptions to this, notably in Australia and New Zealand. (See Chapter 11 for a further explanation.)

Sheep in a Developed Agriculture

In a developed agriculture, the use of sheep differs depending on whether the predominant production system is intensive farm flocks, or extensive range production. Range production, in turn, differs depending on whether the range is dry or wet. In most farm flock situations, meat production is more profitable than wool production. Therefore, when forage conditions are good enough to permit weaning a high percentage of lambs fat enough for slaughter, this will be the primary production goal. Wool production will be secondary and more of a by-product. Wool production is more profitable if forage conditions are poor. When this is the case, more profit will be earned if wool production is emphasized with meat production secondary. This is the situation in the dry range areas, but not in the wet range areas. (Australia vs. New Zealand). The wet range areas are a special situation. They produce enough forage to permit weaning a high percent of the lamb crop fat enough for slaughter so they use dual-purpose sheep.

Sheep in Subsistence Agriculture

The individual engaged in subsistence agriculture is concerned with producing for his own consumption and not in producing a large surplus for sale. Thus, he favors a triple purpose sheep that produces meat, wool and milk. In some subsistence systems the milk is very important. Almost all the Persian Lamb Skins in the world are produced in countries with a subsistence agriculture. They are often an important export item for the country.

Sheep in Primitive Agriculture

In a primitive agriculture, sheep are much more valuable as a source of food than they are for any other use. They are valued more for meat than for milk. In societies where sheep are milked, the milk is usually secondary to meat as a food contribution of the sheep. The lower the level of economic development, the more likely it is that sheep (and goats) rather than cattle will be the farm animal raised primarily for meat.

Wool is not particularly emphasized in primitive agriculture. It is used in cold climate primitive agriculture. However, a disproportionately large portion of the primitive areas of the world are hot and wool is not needed. As a result, many of the strains of sheep found in primitive agriculture are hair sheep. They are a source of skins, which often is a chief export item obtained from the sheep industry in the country. When wool is produced it is usually a coarse, low grading fleece which is used locally.

Goats in Primitive and Subsistence Agriculture

Goats tend to be most numerous in the areas of subsistence or primitive agriculture where they are kept primarily for meat **(Table 2).** There is no developed country in the top ten in either numbers of goats or production of goat meat. Developing countries account for over 90 percent of the world total in both goats and meat from goats. Goats produce only 1.5 percent of the total milk of the world. Only in Africa (10.9% of total milk) and Asia (4.0% of total milk) does the goat contribute more than 1 percent of the total milk.

Goats in Developed Agriculture

Goats are seldom kept in any numbers in areas of the world with a progressive, highly developed agriculture. In the rare case when they are kept in such areas they are virtually never kept primarily to produce meat. They will be kept for milk and mohair. The milk and/or fiber-producing goat may be salvaged by slaughter at the end of its useful life. The meat is not usually sold as goat meat per se, but rather is blended into sausage products. Also, when produced in the developed countries, goats tend to be kept only in areas that are suited only for goats. Remember, in the developed countries cattle and sheep compete for the available grasslands. Goats are considered only when neither cattle or sheep are well adapted.

When goats are kept in developed countries the product they are kept to produce is determined by the amount of available forage. If the area has ample forage goats are kept for milk. If the forage is scarce they are kept to produce mohair. Goats are not kept on any sizable scale in the United States to produce milk. There are a number of profitable goat milk enterprises, but they are very insignificant when compared to dairy cows. In Europe, they are only slightly more important, and are usually kept as a commercial enterprise only if cows cannot be used, such as in very mountainous areas. In fact, most of the improved breeds of dairy goats originated in the alpine regions of Europe.

Two developed countries, the United States and South Africa, have a sizable commercial industry based on the Angora goat and mohair. The U.S. was the worlds leading producer of mohair as recently as 1975. The U.S. industry is centered in southwest Texas where over 90 percent of the total goats in the country are found. In contrast to mohair production, the raising of Kasmir goats and the production of cashmere is restricted to northern Asia. The yield per animal ranges from a few ounces to a pound, and requires considerable hand labor to comb from the goats.

Conclusion

The goat is a valuable agricultural animal under adverse conditions. It will seek out and eat a wider variety of plants than other species and thus makes better use of land. It will eat plants that are refused by both cattle and sheep. In harsh climates the goat will more likely be used for meat with the other uses predominately in milder conditions The lower the level of agricultural development, the more likely it is that goats will be found. Sheep are found in developed agricultures more than goats are and produce more products which are useful. There are more sheep in temperate zones and more goats in the topics.

3903Chap10.doc.sum96

References

Briggs, H.M. Modern Breeds of Livestock. MacMillan, New York.

Bushman, D.H. 1977. The role of poultry production on small farms. J. Animal Sci. 45:402.

Epstein, H. 1969. Domestic Animal of China. Commonwealth Agricultural Bureaux. Farnham Royal, Bucks, England.

FAO 1994. Production Yearbook.

Mason, I.L. 1969. A World Dictionary of Livestock Breeds Types and Varieties. Commonwealth Agricultural Bureaux. Farnham Royal, Bucks, England.

USDA. 1976. Foreign Agricultural Service. Foreign Agricultural Circular FW 2-76.

USDA. 1981. Foreign Agricultural Service. Foreign Agricultural Circular EPE 4-81.

USDA. 1981. Foreign Agricultural Service. Foreign Agricultural Circular FW 1-81.

Taylor, R.E. 1995. Scientific Farm Animal Production, 5th ed., Prentice-Hall.

Study Questions

1. Compare and contrast the goat and sheep populations between temperate and tropical climates.
2. Where are most of the world's sheep found?
3. Where are most of the world's goats found?
4. Describe the various fibers which sheep have.
5. What is the single most important thing determining where there will be large concentrations of either cattle or sheep?
6. What is the difference in the products from sheep in wet range compared to dry range?
7. Why are triple purpose sheep used in subsistence agriculture?
8. How are sheep used in primitive agriculture?
9. Compare and contrast how goats are used in the 3 types of agricultures.

Chapter 11

Cattle Versus Sheep Versus Goats

Introduction

Cattle and sheep are the two most numerous and widespread agricultural animals throughout the world excluding poultry. Goats are the 4th most numerous. How do people choose which one? All 3 require cheap roughages to be produced economically and all provide multiple products/services.

General Considerations

In a developed agriculture, cattle and sheep are almost always the species that are competitors for the roughage (primary grasslands) of an area. Goats are rarely if ever considered. In most developing countries, cattle and sheep are usually preferred, but goats are most numerous. The lower the level of agricultural development the more favored the goat becomes. If both cattle and sheep are equally well adapted, cattle are usually chosen for two reasons: (1) cattle are larger in body size and, therefore, more useful since they can be used for both work and food, and (2) cattle generally require less care and attention. In many parts of the world, cattle were chosen centuries ago and have become the traditional agricultural animals of the area. In many places they have become part of the culture and/or religion of the country.

Under 2 conditions cattle are better adapted than sheep. These are hot, humid climate and when predators are a problem. In such areas, the livestock producer has no choice as to which species he will produce. Under 3 conditions sheep are better adapted than cattle, very dry climates, high altitudes and where there is grass that grows close to the ground. The mouth parts of the sheep permit it to graze such grass more readily than can cattle.

Sheep are also usually the species chosen if herding is necessary because of their flocking instinct. Sheep with a flocking instinct tend to remain together in a compact group. Some breeds have been highly selected for this. Because of this instinct, one person with a good sheep dog or two, can herd a large sheep flock, even in predator country.

Sheep are <u>not</u> very numerous in the humid tropics for 2 reasons. They are not very tolerant of hot, humid conditions. They are also very susceptible to internal parasites which thrive under humid tropical conditions.

Goats usually compete with sheep rather than with cattle for the available forage of an area. In a developed agriculture, the choice is virtually always sheep. The more primitive the agriculture, the more likely goats and sheep are to be equally favored and exist in about equal numbers. In fact, most of the goats of the world are found in areas with subsistence or primitive agriculture. Goats are not as well adapted to high altitudes as are sheep. Goats are better adapted than are sheep to tropical conditions and both are better adapted than cattle to dry conditions with the goat somewhat better adapted than sheep to dry conditions.

In developing countries with some very dry areas, as one moves from the wetter to the very dry areas a very distinct change in the composition of the livestock population is noted. In the wet areas, cattle predominate although sheep and goats are raised. As the country becomes drier, the numbers of cattle decrease, finally disappearing altogether in favor of sheep and goats. Finally, the limit of survival of sheep is reached and in the very dry areas, only goats are found.

Goats have been widely criticized and accused of causing desertification in arid regions. Both sheep and goats with their close grazing ability will severely weaken grasses during a period of stress such as drought, if they are stocked too heavily. The small hooves of both loosen the crust and hasten and accelerate wind erosion. Goats get most of the blame because in the dry regions they are most apt to be the predominant species. The major problem is not the goat, but man's mismanagement of the goat. When properly managed, goats are often the most productive species of agricultural animals for many arid regions.

Subsistence and Primitive Agriculture

Since many of the world's animal-using primitive agricultural systems are found in the dry parts of the world, we should expect to find more sheep and goats with emphasis more on goats the drier it gets. In general this is true. However, tradition and culture are also important determinants and in selected areas have overridden the practical considerations. For example, in much of Africa, goats and sheep would contribute more to the nutritional well-being of the people than cattle. Yet, instead of sheep or goats, cattle are kept in large numbers strictly for the prestige that large herds of cattle accrue to their owners.

There are some high subsistence (approaching developed) agricultures in South America that make extensive use of the sheep and produce large volumes of wool for export. The sheep has been a boon to the economics of these countries for years because of the hard currency the wool brings into these nations. Some of them have turned more to cattle in recent years but the sheep remains important.

In dry, hot, primitive agriculture we find some very interesting adaptations of sheep. Both fat-tailed and fat-rumped sheep are found there. Fat storage serves two major purposes for the body: energy storage and insulation. In these hot areas insulation is little needed but the energy storage is vital to survival. Most tropically adapted animals resort to some type of localized fat storage. In the sheep this is a fat tail or a fat rump. The tail of one of these sheep may weigh several pounds. The two major fur-bearing breeds of sheep are also found in these environments.

Developed Agriculture

In a developed agriculture beef cattle and sheep most frequently compete for roughage sources. Cheap roughage dictates where either is found. This is true for all available roughage, not just the grasslands. This is not true for dairy cattle for reasons discussed later. Profit is the most important deciding factor in the choice of animal species in a developed agriculture. To maximize profit a livestock producer must increase gross returns and/or reduce costs of production. If either of these factors is fixed the other offers the only means of increasing profits.

Sheep or Cattle? Many of the economic factors in the production of beef and sheep are the same in a developed country. The economic returns to each is generally low and fixed. This is difficult to change very much so maximum profit in these enterprises requires cutting production costs. Practically speaking, this means reducing feed and labor costs. Cheap roughage is the primary way of reducing feed costs. We get it 2 places primarily: crop residues from cultivated lands and grass from land which cannot be profitably cultivated such as dry land grassland, high elevations or rough land. The 2 most common ways to minimize labor costs are using extensive management (this means lots of animals and few people) and using animals as a secondary enterprise (this means the animals are worked only when there is no other demand for the labor).

Which agricultural systems use beef and sheep? From earlier studies we have already established this to be Ranching, Livestock Fattening, Commercial Dairy, and Livestock and Crop Farming for a developed agriculture.

Ranching (range livestock production). Its characteristics make it well suited to beef and sheep production because:

- Land can't be profitably farmed for some reason. Thus, the roughage growing on that land is inexpensive because there is no alternative use for the land.
- The land usually has a low carrying capacity per acre.
- Labor cost can be minimized by using extensive management.
- Note: Remember that on a ranch the animals are the primary enterprise and a given ranch usually has either sheep or cattle but not both.

Livestock and crop farming. The animals are referred to as a farm herd. Crop residues are fed to reduce feed costs. Good crop land may be used to produce forage for beef cattle or sheep. If so, the yield must be so good that forage is still cheap. Labor costs are reduced by making the animal enterprise secondary. The animals get attention only when everything else is done and in the off season for crops. Farms often have multiple species such as hogs, cattle and sheep all at the same time. Most beef cattle and sheep enterprises are extensive. However, cattle and sheep enterprises can be intensive. Purebred flocks, herds, feedlots and hobby enterprises are good examples of intensively managed beef and sheep units.

Livestock fattening. This new farming specialization is used for both beef and lambs. The animals are the primary enterprise.

Dairy cattle in a developed economy. Dairy herds have high fixed costs so they must maximize gross returns. This is done by obtaining maximum production through good management and feeding programs. Management must be very good. The management is intensive and very specialized. Dairy feeding programs are based on optimum use of quality forages plus the feeding of supplemental high energy and high protein feeds. The biggest factor in determining where dairies are in a developed agriculture is the location of the market for the milk. Transportation for long distances is cost prohibitive. The two most common markets are near population centers and near manufacturing centers. Examples: Wisconsin, Minnesota, New York, Texas and California have lots of dairy cows. Oklahoma does not.

Conclusion

Cattle, sheep and goats are all widely used agricultural animals. They are used in many types of agriculture and provide a wide variety of products. Which is chosen by a given group of people is dependent on a variety of things. These include overall agricultural system, climate, cultural and social needs as well as the animal's ability to adapt to the conditions in which it is to be grown.

Study Questions

1. Where do cattle, sheep and goats rank in terms of total numbers, compared to other agricultural animals?
2. What influence does the level of agricultural development have on whether cattle, sheep or goats are chosen?
3. How does the level of needed care influence whether cattle, sheep or goats are chosen as the major grazing animal?
4. Under what conditions are cattle better adapted than sheep? . . . sheep than cattle?
5. Why aren't many sheep found in the humid tropics?
6. To what conditions are goats better adapted than sheep?
7. Why can goats be more of a problem than cattle in desertification of arid lands?
8. What are the major deciding factors in a developed agriculture as to whether cattle or sheep are chosen?
9. In a developed agriculture, what conditions make ranching well suited to beef or sheep?
10. How are sheep and cattle used in livestock and crop farming systems?
11. How are dairy cattle used in a developed economy?

Chapter 12

Swine

Introduction

The pig is extremely important from an economic and food-supply perspective. It produces meat (often by scavenging on its own), the hide is used for food and leather, its hair is used for brushes and hundreds of medicinal products are derived from it. Yet, for all its utility, the pig suffers a tremendous image problem. Pigs are depicted as unclean, both literally (dirty as a pig) and figuratively (religious perspective). However, nothing could be further from the truth. In fact, pigs are quite clean if given a chance. They can even be house broken. While it is true that pigs do "wallow in the mud", it is only to keep cool. They have very poor body temperature regulating mechanisms and wallow in whatever is cool and available. If the environment is cool and shaded, they will avoid the mud. If mud and clean water are both available, they will choose the clean water. The problem of being unclean to Jews and Muslims is equally problematic from a practical perceptive. The number of diseases which may be transmitted from pigs to humans is no greater than that which may be transmitted to humans from other animals. It is quite likely that the reason pigs aren't used by those peoples lies rooted in the desert, nomadic past of the people with whom those religions originated rather than in some great fault of the pig.

Pigs are monogastric and are omnivorous which lets them occupy a different niche than herbivorous animals. They are thus found in very different types of production systems. When used in conjunction with ruminants they can utilize different resources and help people better use what is available to them.

Swine rank third among the farm animals of the world in both numbers and distribution. The extent of their distribution is limited somewhat by the fact that they require high energy feeds. Thus, they are poorly adapted to dry regions and for use by nomadic peoples (especially in hot and dry regions). Also, the numbers of swine are greatly reduced in countries where the predominant religion is either Islam or Judaism. For these reasons, there are many countries in which swine numbers are very restricted or almost completely absent. The primary reason for which swine are raised in all levels of agricultural development is for food. The major difference in swine production in the different levels of agricultural development is in the method employed to raise the pigs to slaughter weight not the use intended for the animals.

Swine have some distinct advantages and equally distinct disadvantages. They are a very efficient converter of grain into edible meat. The conversion is roughly 1 pound of meat for 3 pounds of grain. However, when fed this way, they are direct competitors with the human for food. Thus, it would appear that they would not be suited for highly populated, underdeveloped areas where having enough food for the human population is a problem. It would seem that man would benefit more by consuming the pound of grain rather than having it converted to 1/3 lb. of meat. Yet it is in these areas that the pig is most numerous. For example, China is a developing country, the most populous in the world. Yet, she has more pigs per capita than the U.S. The reason is that they also very efficiently convert human garbage into edible food. When fed this way the pig grows very slowly, but eventually produces a high quality human food out of what would otherwise be wasted. In this way, they do not compete with the human population. Rather they complement efforts to get maximum benefit to man of limited food resources. The pig has been a scavenger of wastes since before it was domesticated.

Swine in Developed Agriculture

Swine enterprises have high fixed costs of production in a developed agriculture, primarily as a result of the necessity to feed high energy rations if they are to be economically grown. Therefore, maximum profit depends upon increasing gross returns by improving efficiency of production. This is best done by reducing labor input and the time required to reach market weight. Both of these are accomplished by utilizing expensive confinement systems and full feeding grain to obtain maximum gains. This requires a large investment in buildings and equipment.

Swine in Subsistence and Primitive Agriculture

The pig is well suited for use as a meat animal in the developing countries for two reasons related to the pig carcass: (1) It yields one of the most valuable fats of any domesticated animal, valuable for cooking and many other domestic uses. (2) The pork carcass can be preserved for long periods of time without refrigeration by packing it in salt or brine and still remain palatable and nutritious. Any meat can be preserved by packing it in salt or brine, but few are very palatable after such treatment.

In poor countries swine are not fed grain but rather are used to convert human waste to an edible product. They are seldom confined but rather run free to scavenge what they can in addition to the garbage they are fed. Thus, there is no investment in feed, buildings and equipment. Gains may be slow but eventually valuable protein is gained for the human diets. As the level of agricultural development improves toward high subsistence to marginally developed, the more likely it is that pigs will be fed some grain in addition to garbage. Also they are usually raised in some type of semi-confinement or at least they are not permitted to roam at will as in primitive societies.

Distribution and Utilization of Swine

The leading countries of the world in hog numbers and the production of pork is presented in **Table 1.** China is clearly the leading nation in numbers of pigs. Not only do they have 46% of all the worlds' pigs, they also have considerably more pigs than the rest of the countries in the top ten combined and they are gaining an even larger percentage of the world's total hogs each year. Their overwhelming superiority in numbers insures that they will also rank first in the production of pork. They do not utilize their swine as efficiently as do the developed world producers and their meat production per animal is considerably less. This is to be expected since a high percentage of the total production in China involves various amounts of feeding of human wastes. It must also be understood that one of the primary products from Chinese hogs is the

Table 1. World Swine Numbers, Production, and Consumption

Country	Rank	No. Swine (1,000 head)	Rank	Production (1,000 MT)	Country	Per Capita Consumption (lb.)[a]
China	1	402,846	1	33,835	Denmark	145
United States	2	57,904	2	8,027	Poland	116
Brazil	3	30,450	7	1,290	Czechoslovakia	113
Russian Federation	4	28,600	4	2,400	Sweden	121
Germany	5	26,044	3	3,550	Austria	112
Poland	6	19,466	6	1,609		
Spain	7	18,188	5	2,125		
Mexico	8	18,000	10	873		
Ukraine	9	15,298	8	910		
Viet Nam	10	15,043	9	907		
Africa	5	21,080		719		
N.C. America	3	93,057		10,356		
S. America	4	49,846		2,321		
Asia	1	482,889		40,041		
Europe	2	167,983		20,747		
Oceania	6	4,910		440		
World total		875,407		78,954	**World Average[b]**	29

[a]Carcass weight.

[b]Estimated by dividing world production by world population.

Sources: USDA,FAS (*World Livestock Situation*). 1994 FAO Production Yearbook. Taylor, 995.p51.

manure which they rely on for fertilizer (Ensminger & Ensminger, 1973). They intentionally feed bulky, fiber-containing feeds which are of low digestibility to pigs to increase the collectable manure. Five of the top ten countries in the production of pork are European countries. A large number of the countries of those ranked 11 - 20 are also European. The US produces the most pork per capita of any country in the world.

Swine are found largely in the temperate zones and in areas of dense populations (Ensminger 1977). Hog numbers fluctuate worldwide especially in developed countries in response to available concentrate feeds. In years of plentiful concentrates, hog numbers go up. Conversely, in years of poor harvests less hogs are fed. In the developing world, hog numbers are largely a function of population and not available concentrates. The primary reason for this is because of the large amounts of garbage fed to pigs in these areas. In contrast, less than 1% of the commercially produced hogs in the developed countries are fed on garbage (kitchen wastes) (Pond & Manor, 1984).

Swine in Asia

Asia ranks first in numbers of swine with approximately 55% of the worlds total. Over 80% of the Asian swine are in China. There are over 100 different breeds and varieties of pigs in China, of which 40 are considered to be economically important. Many of these strains are extremely prolific, but few have spread outside of China. Recently some have been brought into the U.S. for experimental crossbreeding purposes. Most of the Chinese breeds are characterized by having very swayed backs with a sagging pendulous belly. The status of these countries as developing ensures that they will continue to keep hogs for their garbage-using and scavenging abilities.

Swine in Europe

In Europe hog numbers are tied closely to two things: the dairy industry and concentrate feeds. Hogs are related to the development of the dairy industry because dairy by-products are important sources of hog feed in some European countries. This includes skim milk, buttermilk and whey. Concentrate feeds for feeding are important to the swine industry of Europe because they need to keep swine in confinement. In the Danube Basin of southern Europe a substantial amount of corn is grown. In this region hog production is a dominant type of farming. Potatoes are raised and fed to hogs in Germany and Poland. In other parts of Europe, excess barley is fed in great quantities to hogs. Swine are extremely important in Europe. As a continent, Europe ranks second behind Asia in total numbers with 19% of the total hogs of the world. This means Europe has 43% of the hogs of the world outside of Asia. Europe ranks second in the production of pork, producing 26% of the total or 53% of the pork produced outside Asia. Five of the top ten countries in the production of pork are European. Several of the top ten countries in export of pork are also European. Europe consistently accounts for approximately 80% of all the pork exports in the world.

Europe is not only a major producer of pork, she is also a major consumer. The top 10 countries of the world in per capita consumption of pork are European. Several European countries cannot produce all of the pork their population needs. Several of the top pork importing countries are also in Europe. There are several reasons why the pig is so popular in Europe. It is the most efficient red meat producing animal for developed, densely-populated countries with surplus grain they desire to convert to meat. The pig is more efficient than either cattle or sheep in feed conversion. The pig is best suited of the large farm animals for rearing in confinement, and land to devote to livestock is usually restricted in densely populated countries.

The Pig in North America

The United States, Canada and Mexico are all in the top 15 swine producing countries in the world. Considering the relatively sparse population of Canada, the many wide open spaces of the US, the developing status of Mexico and the wide climate differences in these 3 countries, this is quite remarkable. Again, the pigs adaptability to a wide range of conditions and its ability to use so many different feedstuffs contribute to this success.

In the United States, swine production has been undergoing a rapid change in structure with large integrated corporate holdings of swine production units rapidly increasing to the exclusion of smaller privately owned units. With the continued improvement in world trade laws opening export markets this trend will continue for the foreseeable future.

Conclusion

The pig is responsible for the production of over 40% of the total world meat production. It is important as a meat producer in virtually every country which has an environment to which it is adapted. This, in spite of its poor image problem. Remarkably, it is the only major domestic red meat-producing mammal that is a monogastric. The rest are ruminants or pseudo-ruminants. Its primary function is as a food producer. In some developing countries (most notably China) it is invaluable as a manure producer.

The pig has largely escaped the social and cultural entanglements that mark many of the world's other major meat producers with the exception of a few religious taboos in religions predominating where the pig isn't very adapted anyway. Only in some primitive Pacific island societies does one find other significant social/cultural involvement. For the most part the pig is not tied up as a religious or status symbol. One can only speculate as to why this is true. Nevertheless, it frees the pig to do what no other red-meat producing species can do: produce red meat very efficiently from a wide variety of feedstuffs on very little land in a relatively trouble-free way and in many, many different physical environments.
3903Chap12.doc.sum96

References

Ensminger, M.E. and Audrey Ensminger. 1973. China - the impossible dream. Agriservices Foundation. Clovis, California.

Ensminger, M.E. 1977. Animal Science. The Interstate Printers and Publishers, Inc. Daneville, Ill. Ch. 36. p. 709.

Epstein, H. 1969. Domestic Animals of China. Commonwealth Agricultural Bureaus. Farnham Royal, Bucks, England.

FAO. 1994. FAO. Production Yearbook.

Mason, I.L. 1969. A World Dictionary of Livestock Breeds Types and Varieties. Commonwealth Agricultural Bureaus. Farnham Royal, Bucks, England.

Pond, W.G. and J.H. Maner. 1984. Swine Production and Nutrition. AVI Publishing Company, Inc. Westport, Connecticut.

Taylor, R.E. 1995. Scientific Farm Animal Production.

Study Questions

1. Does the pig get a bum reputation with regard to the "unclean" issues?
2. What type digestive tract does the pig have? What are the implications of this in determining which niche they occupy in agriculture.
3. Where do swine rank in numbers of animals and distribution?
4. Which religions forbid consumption of pig meat?
5. What is the conversion ratio of feed to food for the pig?
6. Why are pigs so important in developing countries?
7. What must be done to obtain maximum profit from a swine enterprise? How is this achieved? What is the most important single factor determining where there will be large numbers of swine in a country with a developed agriculture?
8. What is the primary reason why swine are kept in the various levels of agricultural development? What is the major difference in swine production in the different levels of agricultural development?
9. What country ranks first in numbers of hogs and in the production of pork? Where does the United States rank in each of these categories?
10. Why is the pig so popular in Europe? In North America?

Chapter 13

Poultry

Introduction

Poultry: We will use the term *poultry* to mean chickens, turkeys, geese, ducks, pigeons, peafowls, guineas and any game birds which are kept as domestic animals. Chickens, ducks, and turkeys are most numerous with chickens being much more important than the other. Ducks are commercially more important in some parts of Asia and there are more geese in some European countries but not enough of either of these to rival chickens on a global scale. There are only about 5.5% as many ducks as chickens and only 2.0% as many turkeys.

Poultry have much in common with swine. They are very efficient converters of grain to food for humans. Unfortunately, when they are the most efficient is when they compete directly with humans for food. They are even better adapted than is the pig for total confinement. This is essentially the only system employed in commercial production of poultry in the developed countries today. They are very efficient in converting human wastes to human food as is the case with swine. Poultry are produced primarily for food as is the case with swine. In areas with a highly developed agriculture, the system of production is very intensive with the birds full-fed a high energy ration for maximum production. In subsistence and lower levels of agricultural development poultry serve as scavengers. In such cases their level of productivity is very low but they do not compete with the human for food.

Distribution and Utilization of Poultry

The numbers of chickens, turkeys and ducks in the world is presented in **Table 1**. Chickens far outnumber either ducks or turkeys. Almost all of the ducks of the world are found in the developing countries. The reverse is true for turkeys. One reason for the duck's popularity in the developing countries is that the duckling is much hardier than are chicks. They require some care the first few days of life, but after that they can survive and grow under conditions that would be disastrous with any other type of poultry. An additional reason for the duck's popularity is that they are excellent scavengers. An unexpected adverse effect of the green revolution has been a reduction in the number of ducks in developing countries. The new high yielding varieties of cereal grains also increased the use of chemical fertilizers, pesticides and herbicides which have all but eliminated the keeping of ducks in many cases, especially in Southeast Asia. The same detrimental effects have been observed in the fish which were once found in irrigation lagoons and canals.

Table 2 gives statistics for individual countries. There are several interesting observations that one can make about the figures in **Table 2**. Notice first that poultry production has increased tremendously in the last 15 years with some countries undergoing phenomenal increases. Many of those experiencing the most rapid growth are Asian. Two things are fueling those increases: 1) burgeoning human populations and, 2) rapid economic development. Chicken is an excellent

Table 1. Livestock Numbers in the World (x1000 head)(1994)

Animal	World Total	Africa	North America	South America	Asia	Europe	Oceania
Cattle	1,288,124	192,180	163,354	279,656	410,118	107,158	34,049
Sheep	1,086,661	208,845	17,529	94,054	340,102	130,692	182,758
Pigs	875,407	21,080	93,056	49,846	482,889	167,983	4,910
Goats	609,488	176,089	14,944	22,819	373,005	14,809	1,007
Buffaloes	148,798	3,200	9	1,436	143,640	133	-
Horses, Asses & Mules	116,882	19,560	20,150	22,049	43,456	5,367	439
Horses	58,158	4,758	12,740	14,493	16,007	4,185	430
Asses	43,772	13,408	3,703	4,105	21,325	906	9
Mules	14,952	1,394	3,707	3,451	6,124	276	-
Camels	18,831	13,815	-	-	4,761	-	-
Total	4,144,194	634,769	309,042	469,860	1,797,971	426,142	223,163

Poultry Numbers in the World (x 1,000,000)

Animal	World Total	Africa	North America	South America	Asia	Europe	Oceania
Chickens	12,002	976	2,086	1,166	5,553	1,154	85
Turkeys	247	6	100	11	11	85	1
Ducks	681	16	15	13	575	41	1
Total	12,930	998	2,201	1,190	6,139	1,280	87

Source FAO Production Yearbook, 1994

Table 2. World Poultry Numbers

Country	Chickens, (1,000,000) 1979-81	1994	1994 as a % of 1979-81	World Rank (1994)	Ducks (1,000,000) 1979-81	1994	1994 as a % of 1979-81	World Rank (1994)	Turkeys (1,000,000) 1979-81	1994	1994 as a % of 1979-81	World Rank (1994)
China	906	2,692	297	1	190	443	233	1	1	1	100	
USA	1,068	1,530	143	2	5	6	120	14	52	88	169	1
Brazil	426	680	160	3	5	9	180	9	2	7	350	6
Indonesia	168	640	381	4	27	27	100	3				
Russian Federation		600		5						23		3
India	160	467	292	6								
Japan	284	324	114	7								
Mexico	177	293	166	8	5	8	160	11	10	6	60	8
France	177	217	123	9	9	20	222	6	14	33	236	2
Iran	97	182	188	10	1	2	200		1	2	200	15
Turkey	55	178	324	11	1	1	100		3	3	100	14
Ukraine		159		12		22		4		1		
Italy	138	137	99	13					18	23	128	4
Thailand	60	127	212	14	12	21	175	5				
UK	116	126	109	15	1	2	200		7	11	157	5
Nigeria	80	122	153	16								
Bangladesh	59	116	197	17	17	15	88	7				
Netherlands	81	110	136	18	1	1	100		1	1	100	
Malaysia	51	98	192	19	8	13	163	8				
Pakistan	44	98	223	20	1	3	300					
Germany	137	96	70	21	1	1	100		1	4	400	10
Canada	96	96	100	22	1	1	100		5	6	120	7
Morocco	24	91	379	23								
Venezuela	42	90	214	24								
Viet Nam	55	85	155	25	28	30	107	2				
Algeria	24	78	325									
Romania	92	77	84		4	5	125	15	2	1	50	
Colombia	30	75	250									
Philippines	53	68	128		5	9	180	10				
Argentina	38	59	155		1	3	300		2	4	200	12
Kazakhstan		55								2		16
Ethiopia	52	54	104									
Spain	51	51	100									
Belarus		45							2	2	100	17
Poland	77	44	57		4	7	175	13	1	1	100	

CONTINUED
Table 2. World Poultry Numbers

Country	Chickens, (1,000,000)				Ducks (1,000,000)				Turkeys (1,000,000)			
	1979-81	1994	1994 as a % of 1979-81	World Rank (1994)	1979-81	1994	1994 as a % of 1979-81	World Rank (1994)	1979-81	1994	1994 as a % of 1979-81	World Rank (1994)
Egypt	28	38	136		3	8	267	12	1	1	100	
Myanmar	23	25	109		4	4	100					
Madagascar	18	23	128		3	5	167		2	4	200	11
Israel	25	23	92						4	4	100	13
Portugal	18	22	122						3	5	167	9
Cambodia	3	11	367		1	4	400					
Asia	2,276	5,553	244	1	292	575	197	1	9	11	122	3T
Europe	1,226	1,154	94	4	25	41	164	2	53	85	160	2
NC America	1,452	2,086	144	2	11	15	136	4	67	100	149	1
S America	686	1,166	170	3	7	13	186	5	4	11	275	3T
Africa	564	976	173	5	9	16	178	3	3	6	200	5
Oceania	58	85	147	6		1		6	1	1	100	6
World Total	7,185	12,002	167		345	681	197		199	247	124	

choice of animal to fill the demand generated for meat. Several Latin American countries are also increasing for the same reasons. Ducks have actually increased at a greater rate during the same period than chickens. In general, the same Asian and South American countries are fueling the increases. Turkeys have never been very well distributed around the world but some countries seem to be developing an improved taste for them. The USA has always dominated this industry but the French and Italians seem to be developing industries. The turkey is less efficient than the chicken and not as hardy as the duck so one would expect its numbers to be greater in developed or the more developed of the developing countries. The numbers in **Table 2** reflect that.

Production of Eggs and Meat

See **Table 3.** China has the highest poultry numbers and ranks second in meat production. The USA has the second highest numbers of animals but leads in meat production. World hen egg production is 39,355,984 MT with the leading countries the same as poultry numbers. Eggs are used for both hatching and human consumption. France and the United States export the largest amounts of poultry, while Germany, Japan, and the former Soviet Union are the leading importers (**Table 4**). Poultry meat is the fastest-growing source of meat for people and production is increasing in all major regions of the world. Developed countries produce and export over 50% of the poultry meat in the world. The United States is the largest producer of poultry meat and the world's leading exporter.

The fact that the developing countries produce less of the poultry meat than the developed does not mean that poultry meat is not important. Poultry products are a valuable source of high quality protein food in developing countries and offer great potential for further improving the nutritional status of these countries. Even extremely poor families can usually obtain a few birds and they are relatively easy to maintain in a limited area. The developed countries also produce the majority of the chicken eggs produced in the world. Egg production in developed countries grew only modestly in the 1980's. However, during the 1990's this appears to be changing. Egg production has been growing in the developing countries at a faster rate than in developed. In 1980, the developed countries were producing twice as many eggs as the developing. Ten years later the ratio is 1.2:1. This is a further testament to the importance of a high quality animal product as being vital to the well-being of low-income peoples. Egg production is increasing in all world regions except Europe.

Poultry in a Developed Agriculture

Poultry are well adapted for confinement and intensive systems of production. When produced in such systems, they offer minimum competition with humans for space and are even more efficient than the pig in converting high energy feeds into quality food. For these reasons, this is the system of production most often practiced in the developed countries However, this is a problem, because when they're produced this way they are in direct competition with humans for food grains. This type of production would be in jeopardy if grain shortages develop.

Table 3. Top 10 Countries of the World in the Production of Hens Eggs and Poultry Meat.

	Hens Eggs			Poultry Meat		
Country	Production MT(1979-81)	Production MT (1994)	Rank	Production 1,000 MT (1979-81)	Production (1000MT) (1994)	Rank
China	2,325,749	10,060,000	1	1,601	6,479	2
USA	4,116,200	4,374,000	2	6,496	13,351	1
Japan	1,998,041	2,562,000	3	1,115	1,320	6
Russian Fed.	-	2,184,000	4	-	1,380	5
India	568,333	1,446,000	5	-	-	-
Brazil	765,117	1,400,000	6	1,345	3,490	3
Mexico	636,256	1,246,223	7	443	1,165	7
Germany	1,123,573	1,100,000	8	-	-	-
France	849,667	982,000	9	1,129	1,898	4
Spain	665,560	669,000	10	-	-	-
UK	834,000	625,832	-		1,085	9
Italy	659,163		-		1,136	8
Canada	330,863		-	532	831	10
Asia	6,962,992	18,227,216	1	5,183	13,790	2
Europe	7,220,058	6,892,153	2	7,017	8,411	3
N. C. America	5,414,971	6,349,092	3	7,789	15,968	1
S. America	1,569,759	2,581,442	4	2,380	6,080	4
Africa	945,466	1,646,930	5	1,100	2,126	5
Oceania	262,725	222,803	6	339	586	6
World Total	26,184,496	39,355,984		25,949	49,125	

Source: FAO. 1994 Production Yearbook.

Table 4. World Poultry Meat Trade.

Exports		Imports	
Country	(mil lb.)[a]	Country	(mil lb.)[a]
United States	1,648	Germany	1,100
France	1,375	Japan	895
Netherlands	979	Hong Kong	686
Brazil	825	Saudi Arabia	550
Thailand	363	Netherlands	374
World Total[b]	7,185	**World Total**[b]	6,019

[a]Ready-to-cook equivalent.
[b]Not all countries included.
Source: USDA, FAS (World Poultry Situation). Taylor, 1995.p43.

Poultry production has undergone a revolution in production technology in the last half of the 20th century. Large integrated companies have come to dominate the poultry industry of the developed countries, especially the USA and Canada. Poultry complexes which contain one million birds at a time are becoming more and more common. These systems have developed because they allow the animals to produce at their optimal level, they protect the animals from environmental hazards, and they are very labor efficient. Often, individual farmers are contracted to raise animals for the company. The companies also make a practice of owning their own production facilities as well.

In the developed countries this type of poultry production is currently serving as the model for other animal industries to become integrated. The swine industry is rapidly moving in the same direction as the poultry industry with other animal industries slowly following suit. The net effect is more efficiently produced and less costly food for the consumer. On the other hand, small farmers find it hard to compete with these large farms and stay in business.

Poultry in Developing Agriculture

Poultry are widely distributed in the underdeveloped countries. They do not compete with the humans for feed in underdeveloped areas but rather use what would otherwise be wasted. In the poorest countries, poultry often exist largely as scavengers finding what they can to eat and are not dependent on their owners even for table scraps. Under such conditions they generally have low productivity, but what they do produce costs little or nothing. Poultry also have an advantage in that they require a low capital investment in birds and equipment. Even the very poor can afford a few birds. Cock fighting is also a very important use for chickens in developing countries. A pleasant afternoon's entertainment might include viewing the cock fight and visiting the concession stand for the barbecued remains of the losers.

Poultry have a great potential for increasing the food supply of developing nations. A developed-country style of poultry production is rapidly expanding into these nations. Most countries undergoing phenomenal increases in production are a direct result of the use of the intensive confinement systems patterned on the developed countries industries. The chicken is a good choice because of its efficiency in feed conversion.

Conclusion

Poultry are very adaptable creatures. They can thrive and produce in conditions ranging from very primitive to very modern and industrialized. Their ability to "live off the land", coupled with their excellent feed efficiencies (better even than pig's), and the fact that they produce 2 valuable foods make them a good choice for many of the world's peoples. The rapid increases in production around the world point to the fact that more and more of the world's peoples are coming to depend on them for high quality food.

References

Briggs, H.M. Modern Breeds of Livestock. MacMillan, New York.
Bushman, D.H. 1977. The role of poultry production on small farms. J. Animal Sci. 45:402.
Damron, W.S. and E.J. Turman. 1991. Agricultural Animals of the World.
Epstein, H. 1969. Domestic Animal of China. Commonwealth Agricultural Bureaux. Farnham Royal, Bucks, England.
FAO. Production Yearbook. Various years.
Mason, I.L. 1969. A World Dictionary of Livestock Breeds Types and Varieties. Commonwealth Agricultural Bureaux. Farnham Royal, Bucks, England.
Taylor, R.E. 1995. Scientific Farm Animal Production. Ch.2.
USDA. 1976. Foreign Agricultural Service. Foreign Agricultural Circular FW 2-76.
USDA. 1981. Foreign Agricultural Service. Foreign Agricultural Circular EPE 4-81.
USDA. 1981. Foreign Agricultural Service. Foreign Agricultural Circular FW 1-81.

Study Questions

1. What are the common poultry species?
2. Where are ducks reared most frequently?
3. How are poultry and swine similar in their production methods?
4. What is the most numerous poultry species?
5. How has the Green Revolution affected ducks?
6. Who are the 3 leading producers of poultry meat and eggs?
7. How has poultry production changed in developed countries in the last half of this century? Contrast this to developing countries.
8. Study **Table 2** and be able to draw some general conclusions about the world poultry industry.

Chapter 14

Horses

Introduction

Worldwide, horses, asses and mules (combined) rank sixth in numbers of the large farm animals **(Table 1).** There are a total of approximately 117 million of the 3 species in the world **(Table 2)**. For comparison, this number is only about 9 percent of the number of cattle. Approximately 3/4 of the horses are found in the developing countries. China has the most of all three species. The top 5 for both donkeys and mules are all developing countries. The horse is a monogastric animal with a functional caecum. This permits it to utilize complex carbohydrates the same as ruminants. Because of this, the horse is adapted to more areas of the world than is the pig, a true monogastric.

Historical

The history of the horse is probably the most interesting of the livestock species because the development of many of the world's great civilizations has been dependent on, or at least assisted by, the horse. Historically, the horse has served man as a beast of burden, a means of transportation, a tool of war, and provided him with recreation, companionship and food. Today, little has really changed. The horse is still doing all those things somewhere in the world.

The horse was one of the last farm animals to be domesticated somewhere between 4500 and 2500 BC. For purposes of comparison, the dates of domestication of other animals were: dog, 12000 B.C.; goat and hog, 7000 B.C.; cattle, 6500 B.C.; and donkey, 3500 B.C. For most of its 5000 years as a domesticated animal, the horse has been primarily associated with three things--war, wealth and nobility. Its use in agriculture and industry on a sizable scale and its ownership by the middle and lower classes has largely been in the last 250 years. During the 19th century and the first 2 decades of the 20th century, the horse became the dominant source of agricultural power. The rapidly expanding agriculture of this period resulted in a doubling of world horse numbers to reach an all time peak of about 110 million in the period 1910-1920. Numbers of horses and mules in the United States peaked in 1920 with 26 million head. The rapid mechanization of farms which began at this time in the developed countries resulted in a rapid decline in the need for horses for farm power and a corresponding decline in horse numbers. In the last 2-3 decades horses have been increasing in numbers because of sport and pleasure use. It has also been increasing in subsistence agriculture as a source of power.

The Use of Horses

The horse has undergone a total change in purpose in the developed world during this century to become a recreational and sport animal. A century ago muscles provided 94% of the worlds energy needs. Coal, oil and water provided the remaining 6%. In developing nations this is still much the same. However, in the developed world we now get 6% of our power from muscles

Table 1. Livestock Numbers in the World (x1000 head)(1994)

Animal	World Total	Africa	North America	South America	Asia	Europe	Oceania
Cattle	1,288,124	192,180	163,354	279,656	410,118	107,158	34,049
Sheep	1,086,661	208,845	17,529	94,054	340,102	130,692	182,758
Pigs	875,407	21,080	93,056	49,846	482,889	167,983	4,910
Goats	609,488	176,089	14,944	22,819	373,005	14,809	1,007
Buffaloes	148,798	3,200	9	1,436	143,640	133	-
Horses, Asses & Mules	116,882	19,560	20,150	22,049	43,456	5,367	439
Horses	58,158[1]	4,758	12,740	14,493	16,007	4,185	430
Asses	43,772	13,408	3,703	4,105	21,325	906	9
Mules	14,952	1,394	3,707	3,451	6,124	276	-
Camels	18,831	13,815	-	-	4,761	-	-
Total	4,144,194	634,769	309,042	469,860	1,797,971	426,142	223,163
				Poultry Numbers in the World (x 1,000,000)			
Chickens	12,002	976	2,086	1,166	5,553	1,154	85
Turkeys	247	6	100	11	11	85	1
Ducks	681	16	15	13	575	41	1
Total	12,930	998	2,201	1,190	6,139	1,280	87

Source FAO Production Yearbook, 1994

[1] This number is greater than the sum of the individual regions. No reason is stated in the source.

Table 2. Top Countries of the World in Numbers of Horses and Production of Horsemeat

		Number of Head				Production of Meat		
Rank	Country	Total Horse Numbers (1000's)	% of World Total Horses	Total Mule Numbers(1000s)	Total Asses Numbers(1000s)	Rank	Total Production (1000 MT	% of World Total
1	China	9,960	17.1	5,498	10,886	3	58	11.6
2	Mexico	6,191	10.6	3,220	3,200	1	80	16.0
3	Brazil	5,800	10.0	2,090	1,370	9	20	4.0
4	USA	3,860	6.6	2	2	5	38	7.6
5	Argentina	3,369	5.8	175	90	4	50	10.0
6	Ethiopia	2,750	4.7	630	5,200			
7	Russian Fed.	2,500	4.3		20			
8	Mongolia	2,100	3.6			6	33	6.6
9	Colombia	2,000	3.4	622	710		5	1.0
10	Kazakhstan	1,400	2.4		40			
	Romania	751			34		8	1.6
	Ukraine	716			15		6	1.2
	Canada	425		4		7	23	4.6
	France	332		13	25	10	12	2.4
	Italy	338		20	30	2	60	12.0
	Australia	272			2	8	22	4.4
	India	990		144	1,600			
	Africa	4,758		1,394	13,408		12	2.4
	NC America	12,740		3,707	3,703		150	29.9
	S America	14,493		3,451	4,105		89	17.8
	Asia	16,007		6,124	21,325		107	21.4
	Europe	4,185		276	906		114	22.8
	Oceania	430			9		23	4.6
	World Total	58,158		14,952	43,772		501	

Source: 1994 FAO Production Yearbook.

(both man and beast) and 94% from coal, oil, natural gas and water power. New uses were found for the horse as it was no longer needed to the same degree as a work animal.

Horses are used in the world today for very different reasons depending on whether the country is developed or developing. In developed countries the use for horses are predominantly for recreation and sport. In developing countries, they are used mostly for transportation and for work. The lower the level of agriculture the more likely we are to find donkeys (asses). The horse is also used for meat and milk in some societies both developed (meat) and developing. In recent years their numbers have increased because of increased use for pleasure and recreation. Since they are still very important as a source of power in underdeveloped countries they have been increasing there.

In the underdeveloped countries, the horse is preferred as a work animal when speed and agility are important. The ox or buffalo is preferred for very heavy work when speed is not an important consideration. The ass is considerably smaller than the horse and, thus can do less work or carry a lighter load. However, it is much more tolerant of hot, arid environments. In hot, dry environments where the agriculture is low level subsistence or primitive, the ass is more likely to be found than the horse. In hot, dry areas with a more highly developed agriculture, the hybrid between the horse and the ass, the mule, is more likely to be used than the ass. A look at **Table 2** confirms this generalization. Notice that Ethiopia has almost twice as many asses as horses and relatively few mules. Mexico has about equal asses and mules which combined equal horses.

Compared to other livestock industries in the developed world, the horse industry has 3 very unique features.

- The horse is not primarily a food species in this country. That makes it very different from all the other species.
- Only a minority of horse owners are actually agriculturists. This is also a large difference.
- Because of reasons 1 and 2, human values enter the horse industry in a way very different from any other livestock industry. The horse business of the developed world is really a people business. The ramifications of this are enormous.

While some horses are still used for work on ranches or in fields in the Developed World, the vast majority of the horses are used for sport or for pleasure. Estimates are that such is the case for between 80 and 90 percent of the horses in the United States, most of them owned by non-farm people. This is not intended to suggest that the production of horses is not an important agricultural enterprise because it is a multi-billion dollar business in the developed countries. However, it does point out the very different perspective that is seen in developed versus developing countries.

For most of its history the horse has been associated with wealth and nobility. For kings and rulers it was, and is, a symbol of pomp and majesty. For many centuries only the wealthy could afford a horse. Thus, to the laboring class it was a symbol of affluence. To some degree, this symbol persists in modern times and the horse, in the minds of many, is one of a long list of status symbols in our modern affluent society. One should be aware that horses serve this same recreation/status role in developing countries, just not as the primary purpose.

Horsemeat

The production of horsemeat is not an important reason for keeping horses. There is no country in which the primary reason for raising horses is to produce meat. The utilization of horses for meat is a salvage situation. It is done when there is no better use for the horse. Therefore, it should not be surprising that the rank of countries in the production of horsemeat is fairly close to their rank in total numbers. Also, horsemeat production increases in poor economic times and times of total herd reductions in the country.

There are several countries that consume a sizable amount of horsemeat. However, it is important to note that, in all cases, horsemeat makes up only a small percentage of their total red meat consumption. The United States exports about 60 percent of its total production of horsemeat. The remainder is utilized in the U.S. as food for pets. Some countries which consume considerable horsemeat (2-4% per capita) are France, Japan, Italy, Netherlands, Belgium, and Poland.

Conclusion

The horse is unique among the major livestock species because the uses it is put to have changed tremendously in the last 50 years or so. Often when uses for a livestock species have become obsolete, the numbers of that species decline dramatically and never recover. The horse followed that same pattern in the first half of this century when the developed countries stopped using them for draft. However, the horse has come to be used much more as a pleasure and recreational animal, assuring its place as a major agricultural animal of the world.

References

Barclay, H.B. 1980. The Role of the Horse in Man's Culture. J.A. Allen. London & New York.

Briggs, H.M. and D.M. Briggs. 1980. Modern Breeds of Livestock, 4th ed. MacMillan, New York.

Evans, J.W., A. Borton, H.F. Hintz and L.D. VanVleck. 1977. The Horse. W.H. Freeman and Co. San Francisco.

FAO Production Yearbook. 1994.

Geddes, C. 1978. The Horse-The Complete Book of Horses and Horsemanship. Octopus Books Ltd. London.

Study Questions

1. Where does the horse rank in numbers as an agricultural animal? Which country has the most?
2. What is the significance of the fact that the horse has a developed cecum?

3. What are the uses of the horse in the world today? How do they compare between developed and developing countries?
4. Describe the conditions under which a horse would be chosen over a buffalo. In the same way compare the ass and horse, horse and mule, ass and mule.
5. During most of its existence as a domesticated animal, the horse has been associated with what 3 things? Approximately how long has it been used on any sizable scale in agriculture and industry?
6. Describe how the horse evolved as a pleasure animal in the USA.
7. Describe the use of horsemeat around the world.

Chapter 15

Buffaloes, Camels, Lamoids and Yaks

Introduction

This chapter has short sections about several species in it. However, the reader should not presume that there is not much information on these species or that they are unimportant to the world's peoples. Nothing could be further from the truth. However, there are fewer of them than the other species who got fuller consideration and they are more restricted in their distributions. One simply has to draw the line somewhere and this is one way to do so.

Water Buffalo

Introduction

The domestic buffalo (Bubalus bubalis) is descended from the wild water buffalo (Bubalis arnee). The domestic buffalo found in different parts of the world are somewhat different from each other. Since they have been subjected to very little selection, then the differences are probably because they are descended from different wild subspecies many of which are extinct.

The water buffalo is found almost exclusively in hot, wet areas with subsistence or primitive levels of agriculture in which rice is produced. They are extremely important in these areas of the world, but are seldom found in any significant numbers as agricultural animals in other parts of the world. They are kept primarily for work, and more specifically for pulling plows. It has enormous strength, a docile temperament and the capacity for long sustained heavy work. They may be utilized to pull wagons or for riding, but their primary use is in field work and much of this in rice paddies. They are a valuable source of milk in Asia which has almost all of the world's buffalo and also contribute substantial quantities of meat.

Although the water buffalo is primarily found in hot, humid areas, it is not very heat tolerant. One reason for the lack of heat tolerance is it has very few sweat glands, only 1/5 the number found in cattle. As a result, it can survive only if it has water in which it can wallow at frequent intervals to give up heat by conduction and to coat the skin with moisture to be evaporated. The two types of domestic water buffalo, swamp and river, derive their name from whether they prefer clean, running water (river type) or will wallow in mud holes (swamp type).

Distribution and Utilization of Buffalo

Table 1 presents the top countries of the world in numbers of buffalo and the production of buffalo meat and milk. Over 96% of the world buffalo population are found in Asia. India is easily the leading country in numbers of buffalo. Add China and over two-thirds of the worlds buffaloes are accounted for. Almost all of the buffalo outside of Asia are found in Egypt and Brazil. A few European countries have some, most notably in Italy. The best mozzarella cheese

Table 1. Top Countries of the World in Numbers of Water Buffalo and the Production of Buffalo Meat and Milk.

		Number of Head		Production of Meat		Production of Milk	
Rank	Country	Total Numbers (1,000's)	% of World Total	Rank	Total Production (1,000 MT)	Rank	Total Production (1,000MT)
1	India	78,825	53.0	1	1,204	1	31,211
2	China	22,416	15.0	3	252	3	2,100
3	Pakistan	18,887	12.7	2	468	2	12,021
4	Thailand	4,257	2.9	7	58		6
5	Indonesia	3,512	2.4	8	52		
6	Nepal	3,176	2.1	5	100	5	631
7	Egypt[1]	3,200	2.2	4	173	4	1,580
8	Viet Nam	3,009	2.0	6	94	10	29
9	Philippines	2,630	1.8	9	44		15
10	Myanmar	2,130	1.4	10	22	8	97
	Brazil	1,435	1.0				
	Turkey	316	0.2		8	6	140
	Iran	300	0.2		10	9	90
	Italy	92	0.1			7	126
2	Africa	3,200			173		1,580
5	NC America	9					
3	S America	1,436					
1	Asia	143,640			2,356		46,470
4	Europe	133			3		140[2]
	Oceania						
World Total		148,798			2,532		48,190

Source: 1994FAO. Production Yearbook.

1 Egypt has almost all the water buffaloes found in Africa.

2 Bulgaria and Italy.

to be had in Italy is made from buffalo milk. It generally sells for around $2-4 more per kilogram than that made from cows milk. In the last 10 years, buffalo herds have increased. Once found primarily in southern Italy, mostly in the countryside surrounding Naples, there are now herds in northern Italy. This is due to the fact that buffalo milk is not controlled by European Community milk quota restrictions (De Luca. 1995).

The water buffalo is not generally raised primarily as a source of meat. Their meat is utilized as a salvage situation through the slaughter of old, crippled, barren or excess animals. India, with over half of the world's buffalo, has traditionally done a poor job of harvesting the buffalo for meat. In recent years this has changed. While it still does not harvest a percentage of meat equal to its animal numbers, more is being harvested. Historically the low utilization of meat was indirectly the result of the religious prohibitions against the slaughter of cattle and consumption of beef. There are no similar prohibitions against the slaughter of buffalo, but the meat is so similar to beef that there are scruples against eating it. Also, for religious reasons many Hindus are vegetarian. The second reason is simply that these animals are more important as draft animals. There is an observable change in this low utilization, however. Meat production has steadily risen in the last two decades and India is now the major producer. With 53% of the animals they produce 47% of the meat. Why this is so is not really known. It may simply be that the increasing population is forcing people to alter their philosophical stand on this issue.

Milk is the second most important product. In Asia they produce 40% of the milk consumed by people. It is 7-8% fat and very nutritious. The rank in amount of milk produced is very similar to animal numbers. It is worth mentioning that Italy is ranked 7th in milk production.

The Advantages of the Buffalo Over Cattle

The buffalo is superior to cattle for work animals in the hot, humid regions where rice is grown in 2 ways: 1) utilizing the coarse, low quality roughages, such as rice straw and 2) working in low, wet areas, such as the rice paddies, where their feet must be in mud and water for long periods of time (cattle are more susceptible to foot rot). Cattle are superior to buffalo in 3 major ways: heat tolerance, particularly to dry heat; as work animals on dry ground because they are much faster, although not capable of pulling loads quite as heavy, and as producers of meat and milk.

Camelidae

Introduction

The family Camelidae were domesticated about 4,000 to 6,000 years ago. Since their domestication, they have provided meat, milk, wool, manure for fuel, offerings to the gods, and hides to the desert- and mountain-dwelling peoples of the Eastern and Western Hemispheres. There are six members of this family. The camel genus has two of them: the one-humped dromedary and the two-humped Bactrian camel. The llama genus includes the guanaco, the llama, and the alpaca, and the vicuna genus has just the vicuna. The camel family is commonly referred to as camelids and members of the llama and vicuna genus are called lamoids.

The Camelidae are very different but they also all share certain characteristics. They all have long necks, small heads and no horns or antlers. Each has a split upper lip similar to that of a rabbit's. The camel is a ruminant but its stomach has only three chambers, rather than the four other ruminants have. It can make good use of plant material because of its ruminant status and they graze many different species of plants. The red blood cells of camelids are oval, whereas in all other mammals they are circular. Their feet have two toes, with a very flat hoof at the end of each. A thick, calloused pad forms the sole of the foot.

Dromedary and Bactrian Camel

Domesticated camelids are primarily found in Africa, Asia, and South America. There are also many camels in Australia because they were used in the Outback as pack animals during the 19th century. Some of the descendants of these camels are feral and are considered to be pests. Since it can go without water for several days (approximately 10), it is a very valuable animal for dry climates. For millennia the camel has helped the desert populations of Asia and Africa survive and even prosper at times. Camelids can travel great distances in desert conditions for days with no water while carrying a heavy load. This ability has earned it the nickname "ship of the desert." It is used mainly for riding or as a pack animal, although occasionally it may be yoked to a plow. It is usually paired with an ox or sometimes an ass rather than another camel because the camel can't be very successfully driven to a plow but the ox or donkey can.

As with other species found in primitive agriculture, the camel supplies many other products in addition to its primary product (work). Although it is very seldom raised primarily for food, camel meat and milk are also put to good use. When it is used for meat, it is usually in a salvage situation such as where animals are not useful for work because they are old or crippled or when the herd needs to be reduced in size. The Somali tribe of Africa does raise camels for meat, but they are the exception rather than the rule. Lactating female camels may be milked and many constitute the major source of milk for many nomadic people. The hair is turned into tents, blankets, rugs, clothing, and rope. Dried dung is used as fuel for cooking fires. The hides are used for harnesses, sandals, water bags, rugs and many other necessary items.

There are two species of camels: The Arabian is the one-humped camel of the Near East and Africa. It is referred to as the dromedary. The Arabian is fast and considered to be the more desirable for riding. However, it is also widely used as a pack animal by the nomadic people of desert and semidesert regions. This kind is well adapted to hot, dry climates. The Bactrian is the two-humped camel of northern China. It is often referred to as the Asian camel. The Bactrian is a larger slower animal. It is primarily used as a pack animal. The camel pack trains that have moved for centuries across the trade routes of the Orient were made up of Bactrian camels. Since much of the region where the Bactrian is used is cold, it has a longer hair coat (1 foot in winter) than the Arabian and has adapted well to cold conditions. It has a much shorter, thicker, and heavier body than the dromedary's. It is better able to stand rocks, snow and ice because it has more calloused feet than the dromedary. It can also drink brackish (salty) water and swim for short distances. The camel's-hair cloth used to make fine overcoats and other clothing comes from Bactrian camels. The most valued hair grows next to the skin.

For centuries, the breeding of camels has been the primary source of income for many nomadic tribes living in the very dry areas of the world. Although the camel is still an important beast of burden, the steady increase in mechanization is reducing the use of camels. This decreased demand for their product threatens the traditional way of life of the nomads. In the not too distant future they may be forced to give up their nomadic existence and settle in villages. This development would be welcomed by most of the countries of the world which have a sizable nomad population. However, it is often true that Nomads make lousy villagers. They tend to make deserts, then starve.

Ill-tempered and irritable, camelids bellow, bite, kick, and spit to demonstrate their annoyance. They seldom work without a protest. The noise from a camel yard when a caravan is being loaded would "wake the dead". Nevertheless, they are very adapted for the work they are asked to do. No other domestic animal can live and work in such a hot and dry climate with so little food and water. A camel can carry from 500 to 600 pounds and travel 25 miles a day. The *mehari* breed of camel is bred for warfare and racing. It can travel 75 to 120 miles a day at a steady trot of 9 to 10 miles an hour.

Camel Features

The camel's hump (or humps) is fat and muscle and has no bone. When a camel is well fed and given enough water, the hump is erect. When it goes without food and water for a period of time, the fat in the hump can nourish it for several days. It is easy to tell if a camel is in this condition because its hump becomes limp and leans to one side.

Long, interlocking eyelashes in a double fringe surround the eyes protecting them from sandstorms and the sun. Its nostrils are also specially adapted. They are slanting slits which the camel can open wide to breathe or close to keep out blowing sand. Another curious adaptation is a groove from each nostril that ends in the split upper lip. This allows any moisture from the nostrils to flow into the mouth helping in water conservation.

The camel can live without water far longer than other mammals. The reason is that the camel conserves water by holding it in tissues and cells rather than using it to cool itself through perspiration as other animals do. As the environmental temperature increases, so does the temperature of the camel's body. During the night the camel gives off the heat stored during the day so that its morning temperature is low. Their body temperature can vary more than 11° F (6° C).

There are approximately 19 million camels in the world. **Table 2** gives the numbers for world regions and major countries. Notice that all are African or Asian. Notice that China's camel numbers have declined significantly in the last decade. This is no doubt due to the rapid economic development of that country.

Table 2. Top Countries in the World in Numbers of Camels

Country or Region	Total Numbers 1979-81, (1,000's)	Total Numbers 1994 (1,000's)	Rank (1994)	1994 as a % of 1979-81
Somalia	5,883	6,000	1	102
Sudan	2,611	2,856	2	109
India	1,050	1,520	3	145
Pakistan	844	1,121	4	133
Ethiopia	980	1,000	5T	102
Mauritania	734	1,000	5T	136
Kenya	608	815	7	134
Mongolia	476	415	8	87
Saudi Arabia	284	415	9	146
China	597	373	10	62
Niger	371	370	11	100
Africa	12,667	13,815	1	109
NC America				
S. America				
Asia	4,140	4,761	2	115
Europe				
Oceania				
World Total	17,042	18,831		110

Guanaco, Llama, Alpaca and Vicuna

Introduction

These are members of the camel family found only in the high elevations of the Andes Mountains in South America. They consist of four species in two genera. *Lama glama*, the llama; *Lama pacos*, the alpaca; *Lama quanacoe*, the guanico; and *Vicugna vicugna*, the vicuna. They will interbreed and do produce fertile hybrids. The most commonly produced hybrids are between the llama and alpaca (known as huarizo) and between the alpaca and vicuna (paco-vicuna).

Because the llama and alpaca are the two domesticated species, they are the most economically important of the llamoids. The llama is the largest, weighing up to 265 lb. It is found in the highland areas from northern Peru to northern Chile and Argentina and is used primarily as a pack animal. The alpaca weighs up to 165 lb. and is raised primarily for its wool, which has an extremely fine fiber grading 80's on the Bradford scale compared to 60's for wool from the Merino sheep, which produces the finest wool of any sheep. The higher the number a fleece grades on the Bradford scale the finer the wool fiber and the more valuable the wool is for spinning and weaving into very expensive woolen material. There are two types of alpacas, the Suri, which produces a long, straight fiber, and the Huacaya, which produces a more desirable, shorter, curly fiber resembling Corriedale type wool.

The guanaco and vicuna are not domesticated though some are sometimes tamed. The vicuna is the smallest of the two (75-100 lb.) but produces a wool with the finest fiber of any animal, grading 120's on the Bradford scale. Because of the demand for this very luxurious wool, the vicuna was hunted almost to extinction since the only way the fleece could be obtained was to kill the animal. Strict governmental regulations have been imposed forbidding the killing of vicunas in an effort to preserve the species.

The blood of the lamoids has more red blood cells per unit volume of blood than other mammals. In addition, the hemoglobin of the blood reacts faster with oxygen. This gives the animals the ability to exert themselves strenuously at high altitudes and makes them well adapted to their environment.

Llama. *(L. glama)*. Llamas are very plentiful in the Andes Mountains from southern Peru to northwestern Argentina. As the largest of the lamoids, it is kept primarily as a pack animal, but is also valuable as a source of food, wool, hides, tallow for candles, dried dung for fuel, and as a provider of offerings to the gods. Llamas can carry loads of up to 210 pounds for up to 16 miles a day. Llama wool varies in color, being various shades of white, brown, red and roan and often has dark brown or black spots. Though woven into a warm cloth, the fleece is coarse and inferior to that of the vicuna and the alpaca.

Alpaca. *(L.pacos)* Smaller than the llama, the alpaca is often described as looking like a large goat with a camel's head and neck. The alpaca is kept and bred for its wool, which in the suri breed can grow long enough to touch the ground. The wool comes in 22 colors from black to tan to white, is lightweight and has good insulation value. It is used in making items like parkas, sleeping bags, and fine coat linings, among many other things.

Yak

Introduction

The yak is a longhaired, very hardy ruminant found in and around the Himalayan Mountains of central Asia. It is an animal adapted to high altitudes and cold climates and is very important to people who live where cattle cannot live. It is primarily used as a pack animal, and is even ridden, being very sure footed and having great strength and stamina. It is used at lower elevations in pasture farming as a draft animal. The yak weighs 400-1200 lbs. and can carry loads of 300 lbs. over steep mountain trails for great distances. The domestic yak is insensitive to cold and requires little food, and is very sure footed. These characteristics make it an excellent pack animal. It is considered to be the best suited domestic animal in Asia for elevations exceeding 2000 meters. Their nickname translates into "Ships of the Plateau" (Porter, 1991). Yaks survive, even thrive in some of the world's harshest climates, even in the cold desert. In northern Tibet (one region where they thrive) the annual mean temperature is 0°C.

Meat and Milk

The milk production is only about 100 gallons per year but is a very important product. The milk has a high nutritional value with a fat content of 5-8% percent and makes excellent butter and cheese. In Tibet a milk powder is prepared (besides butter and cheese) by a special process of coagulation and is used for provisions. As with many other classes of work animals, generally only the old animals are slaughtered. The meat is often dried and then eaten without being cooked. In Tibet there is a taboo on killing them for meat, however, in Mongolia and several of the former USSR states (now independent) there seems to be a taste for the dark rather coarse meat with its unique flavor.

The yaks are sheared or plucked once a year. The hair is a major export item. They yield about 1600 grams of hair and 600 grams of coarse wool which is spun into yarn. The yarn is made into blankets, tent covers, halters, bags, sling shots, and ropes. Garments are made from the soft wool of young animals up to the age of 3 years. Apparently the fur on the hump and the shoulders is best for this. The underwool is made into felt.

Tibetans also bleed the yak occasionally. The blood is mixed with salt and water and allowed to clot. It is then fried or boiled or sometimes mixed with barley flour and butter and made into bread (Larrick and Burch, 1986). Dried yak dung is collected and used for fuel for home fires. In the treeless area where it is produced, firewood is almost nonexistent. The tail is a prize product being used as fly wisks, religious symbols and cap decorations.

The yak belongs to the genus Bos, species grunniens (grunting ox). It belongs to the Bisontine sub-group which also includes the American bison. The yak will interbreed with cattle (resulting in a yakow) but the male calves that result are not fertile and female hybrids must be bred back to a male of one or the other of the parent species. In the frontier area of Tibet and in some of the countries of the former Soviet Union, yaks are often crossbred with cattle. The crossbreeds are in great demand, since they are especially strong and good natured animals for plowing and use in caravans and they produce more meat and milk. The yakow does better in warmer temperatures and lower altitudes than the full-blood yak though they don't do as well at higher elevations and lower temperatures. Crosses between yak and zebu cattle also give more milk and better meat and are excellent work animals for carrying loads at moderate altitudes. The yak has even been crossbred with the gayal, American Bison, European Bison and Banteng. In general, there is much hybrid vigor displayed in the offspring.

Most yaks are found in Tibet, Western China, Tajikistan, Kirgizizia to Mongolia and Siberia with 85% of the total in China. World numbers are estimated at 13 million head. It is hard to get an accurate estimate because they are often counted as cattle.

References

Bushman, D.H. 1977. The role of poultry production on small farms. J. Animal Sci. 45:402.

Compton's Interactive Encyclopedia. 1995. Compton's New Media, Inc.

De Luca, Giovani. 1995. Personal Communication. pef0655@comune.bologna.it

Epstein, H. 1969. Domestic Animal of China. Commonwealth Agricultural Bureaux. Farnham Royal, Bucks, England.

Grzimek, Bernard. Grzimek's Animal Life Encyclopedia.

Harlan, J.R. 1976. The Plants and Animals that Nourish Man. Scientific American:3,CCXXXV.

Larrick and Burch. 1986.

Mason, I.L. 1969. A World Dictionary of Livestock Breeds Types and Varieties. Commonwealth Agricultural Bureaux. Farnham Royal, Bucks, England.

Porter, V. 1991. Cattle-A Handbook to the Breeds of the World. Facts On File, New York.

Taylor, R.E. 1995. Scientific Farm Animal Production. Chapter 1,2. Prentice-Hall.

USDA. 1976. Foreign Agricultural Service. Foreign Agricultural Circular FW 2-76.

USDA. 1981. Foreign Agricultural Service. Foreign Agricultural Circular EPE 4-81.

USDA. 1981. Foreign Agricultural Service. Foreign Agricultural Circular FW 1-81.

Wunschmann, A. 1968. The Wild and Domestic Oxen in Grzimek's Animal Life Encyclopedia. Volume 13. Mammals pp. 331-398.

Study Questions

1. What country ranks first in numbers of water buffalo?
2. Which ranks second?
3. Which country ranks first in production of meat from water buffalo?
4. What has been happening to the use of buffalo for meat in India over the last Decade?
5. What tie does the religious prohibition on killing cattle have to do with slaughtering water buffalo in India?

6. What European country has a small water buffalo industry?
7. What part of the world uses water buffalo milk in a major way?
8. Describe the advantages the buffalo has over cattle.
9. Who are the members of the family Camelidae? What are the individuals used for by the worlds people and where?
10. Describe the camel's temperament.
11. Why is it that the camel can go so long without water? (All the reasons.)
12. How many camels are in the world and where are they?
13. What are the genus and species of the lamoids and their common names?
14. Compare wool quality between lamoids.
15. What is odd about lamoid blood?
16. Describe the alpaca.
17. Where is the yak found? What does it do?
18. What are the strengths of the yak?
19. How much yak milk is produced each year? What are its characteristics?
20. How often is the yak sheared and/or plucked? What is the wool used for?
21. What are all the uses people make of yaks and their products?
22. What are the advantages of the yakow?